As temperatures continue to rise and we experience a greater frequency of extreme weather, drought, and over-extraction of groundwater, effective water policy management is at a critical juncture. Farley, Fisk, and Morris explore these complexities through the study of three exceedingly divergent states: Alabama, California, and Texas. The approach highlights how water policy is driven as much by political culture, institutional characteristics, and historical context as it is by need and best practices. This work will unquestionably be of interest to policymakers and scholars interested in better understanding the complexities and challenges of state water policy.

  —Martin Mayer, *University of North Carolina at Pembroke*

The Drought Dilemma stands out as a pivotal and enlightening read. This book skillfully navigates the challenging water drought policies in Alabama, Texas, and California, offering key insights into one of our time's most pressing environmental challenges. Detailed interviews enhance the authors' extensive research, illuminating the multifaceted aspects of drought management and water policy. This work reveals that the complexities of environmental problems and policy innovations defy simple categorizations like political ideology or regional location. By bringing the critical aspect of environmental resource distribution to the forefront, the book contributes significantly to our understanding of policy variations and their broader implications and serves as a vital guide for understanding and improving environmental policy. It's an essential addition to the field, poised to influence both current discussions and future policy developments in water management. This book will make a valuable contribution to environmental and natural resource literature.

  —David P. Adams, *California State University at Fullerton*

The Drought Dilemma presents an examination of water policy innovation through a comparative case study analysis across three states: California, Texas, and Alabama. This work underscores the dynamic nature of water problems in these states and the growing challenges facing state and local governments to address regional water issues through effective water management approaches. The central message, the efficacy of water management, requires innovative policy solutions that adapt to changing environmental conditions and mediate future water crises. This study offers environmental scholars, water management practitioners, and students interested in natural resource policy a unique framework to understand the complexities of state-level water policies and the importance of governmental response to meet current and future water needs.

  —Luisa M. Diaz-Kope, *University of North Georgia*

The Drought Dilemma

Water policy in United States is one of the most complex topics in the field of public policy. This book, a comparative study of Texas, California, and Alabama's drought response, provides for the first time a common framework for analysis to investigate how water scarcity and droughts have interacted with various state-level factors to produce a wide degree of variance in policy innovations. Using Toddi Steelman's (2010) conceptual framework, the authors examine multiple variables that impact water policy innovation, while showing how one policy solution does not fit all. They expertly demonstrate divergence in water policies due to the environmental cultures, water distribution, and structures in each case, despite similar drought conditions.

As water is increasingly stressed in the future, the ability to draw on lessons learned by these states will provide valuable insight to other entities that face droughts and water shortages. *The Drought Dilemma* is a must read for all those looking for recommendations for the construction of drought policy, as well as future approaches to understand comparative state drought policy.

Jonathan D. Farley is a contract simulator instructor at Naval Air Station Meridian, MS. He retired from active duty in the United States Navy after a 20-year career as an F/A-18 pilot. His research interests include environmental management and international resource conflict. He has published on water conflict in Asia and future military concepts, such as multi-domain operations. He previously served as an instructor at the Air Command and Staff College located at Maxwell Air Force Base, AL. In addition to his role as an instructor, he was a guest lecturer at the Royal Danish Defense College, German Bundeswehr Command and Staff College, and the NATO School Oberammergau.

Jonathan M. Fisk is an associate professor in the Department of Political Science at Auburn University. He has published nearly two dozen single/co-authored articles/book chapters in journals including the *American Review of Public Administration*; *Energy Policy*; *State and Local Government Review*; *Risk, Hazards, and Crises in Public Policy*; *Public Integrity*; *Society* *and Natural Resources*; *Politics and Policy*; and *Review of Policy Research.* He has worked on grants from the National Science Foundation, the Environmental Protection Agency, and the Alfred P. Sloan Foundation.

John C. Morris is a professor in the Department of Political Science at Auburn University. He has studied environmental policy and water policy for more than 30 years and published widely on public administration and public policy. He is the author/coauthor/editor of 12 books, more than 85 scholarly articles, and more than 40 book chapters and reports.

Routledge Research in Environmental Policy and Politics

The Drought Dilemma

States, Innovation, and the Politics
of Water Quantity

**Jonathan D. Farley, Jonathan M. Fisk
and John C. Morris**

NEW YORK AND LONDON

First published 2024
by Routledge
605 Third Avenue, New York, NY 10158

and by Routledge
4 Park Square, Milton Park, Abingdon, Oxon, OX14 4RN

Routledge is an imprint of the Taylor & Francis Group, an informa business

ISBN: 978-1-032-76137-4 (hbk)
ISBN: 978-1-032-81188-8 (pbk)
ISBN: 978-1-003-49853-7 (ebk)

DOI: 10.4324/9781003498537

Typeset in Times New Roman
by Apex CoVantage, LLC

To Melissa, Layla, Emma, Jackson, and Mason

—Jonathan D. Farley

To Melanie, Noah, Caleb, and Micah

—Jonathan M. Fisk

To Molly and Chris

—John C. Morris

Contents

Figures

Tables

Acknowledgments

We would like to express our appreciation for the state officials in our case study states who were kind enough to share their thoughts and expertise with us for this study. Their willingness to talk with us made this study possible, and we are indebted to each of them. We also owe a debt of gratitude to Steven P. McKnight, PhD candidate at Auburn University, for his assistance in the preparation of this manuscript. Finally, we thank the chair of the Department of Political Science at Auburn, Dr. Cathleen Erwin, for her support for this project.

Abbreviations

ADAPT	Alabama Drought Assessment and Planning Team
ADECA	Alabama Department of Economic and Community Affairs
ADEM	Alabama Department of Environmental Management
AEMC	Alabama Environmental Management Commission
AF	Acre-Feet (approximately 326,000 gallons of water)
ARA	Alabama River Alliance
AWAWG	Alabama Water Agencies Working Group
AWRC	Alabama Water Resources Commission
CalEPA	California Environmental Protection Agency
CVP	Central Valley Project (California)
CWP	California Water Plan
DWR	Department of Water Resources (California)
ESA	Endangered Species Act
GSA	Groundwater Sustainability Agencies (California)
GCD	Groundwater Conservation Districts (Texas)
MWD	Metropolitan Water District of Southern California
NOAA	National Oceanic and Atmospheric Administration
OWR	Office of Water Resources (Alabama)
SGMA	Sustainable Groundwater Management Act (California)
SWB	State Water Control Board (California)
SWP	State Water Plan (Texas)
SWP	State Water Project (California)
TAMU	Texas A&M University
TCEQ	Texas Council on Environmental Quality
TWC	Texas Water Code
TWDB	Texas Water Development Board
WWD	Westlands Water District (California)

1 State Policy and Droughts

There are few necessities in life, but water is one that all humans, regardless of culture, politics, goals, or location, have in common. The socio-ecological challenges of water scarcity, primarily driven by the confluence of climate change, hyper-urbanization, mismanagement, growing demand, and pollution, are estimated to affect almost one half of the world's population annually, including much of the American West (Mekonnen and Hoekstra 2016). Droughts have historically impacted multiple Western states, but as populations have grown in this region—currently 20 percent of the U.S. population lives within California and Texas alone—the number of Americans at risk from severe drought has increased. Similarly, as populations swell around the world, providing water in the face of increased demand and increasingly severe and frequent droughts will place additional pressures on a variety of stakeholders. Given these challenges, the question then becomes: how do states with varied water resources, distinct institutional development, and diverse political cultures react to drought events?

Within the United States, water policy is among the most complex topics in the field of environmental policy. From the politically charged term "climate change" to the tensions between and among states and substate entities to the cacophony of thousands of interest groups, water policy is often combative. Moreover, the diversity of the United States creates unique problems for the students of water policy and planning, as the rate, intensity, and severity of droughts vary dramatically. Despite these challenges, water planning is a necessary part of environmental governance, quality of life, and economic development. As such, this book demonstrates that innovation is possible, that multiple variables impact water policy, and that one size does not fit all. As water is increasingly stressed in the future, the ability to generate and identify lessons learned will provide valuable insights to other entities that face droughts and water shortages. In short, understanding the factors that contribute to different water management solutions, both successful and unsuccessful, will be invaluable.

DOI: 10.4324/9781003498537-1

Goals of the Book

To address the goal of understanding state-level water quantity innovation, we examine drought responses in Texas, California, and Alabama's using Steelman's (2010) policy innovation and implementation framework. These states differ dramatically relative to their dominant political ideologies, water availability, and drought responses, but each has faced serious drought in recent years; we turn to each next:

- From 1950 to 1957, the state of Texas experienced a drought that forever changed the rural landscape. Lasting over 71 months, this drought reshaped the Texas population, shifting from a largely agrarian and ranching culture to a predominately urban culture. In short, this "drought of record" left all but one of the 254 counties classified as disaster areas and is the foundation of all modern water conservation within the state (TWDB 2012, 2017, 2019).
- Tree ring analysis in California points toward a history of century-long, catastrophic droughts in the region (Lund et al. 2018). California's drought from 2012 to 2016 was a once in every 1,200-year drought, killing an estimated 102 million trees (Lund et al. 2018), years' worth of salmon hatcheries (Bland 2015), and resulted in groundwater depletion equivalent to about one-third of the total volume of Lake Tahoe (Friedlander 2018).
- In contrast, the state of Alabama is one of the most water-rich locations in North America; but, despite large rivers that crisscross the state and normally abundant rainfall, drought has still affected this water-rich region. The most recent drought, in 2016, killed tens of thousands of trees across the state, which resulted in over 50,000 acres of land burned from wildfires, and left many creeks and rivers dry enough to walk across (Alabama Forestry Commission 2016).

Within these states, we investigate how environmental culture, historical decision-making, contemporary institutional structures, and rules, as well as the geographic distribution of water resources, have contributed to differences in water policy innovation.

The use of Steelman's (2010) model as a guiding framework provides a useful lens to examine water policy innovation as it was developed to address the peculiarities of environmental policy innovations, via a wide swatch of cultural, structural, and individual factors (18–19). Additionally, the framework allows in-depth research across and between units of government and sectors. Finally, the model is versatile and can accommodate a range of phenomena ranging from external and episodic events to more subtle and more technical policy shifts (Kingdon 1995; Birkland 1997, 2006). Despite drought and climate change research, comparative case studies of water policies are rare. Of the examples in the literature, most compare legal structures and policies across international cases (Meinzen-Dick 2007; Dingfelder 2017; Persons 2017). This comparative analysis will provide the first known multivariate test of Steelman's framework through three states, spanning environmental and political spectrums.

The Wicked Nature of Drought

Droughts and water quantity represent a wicked problem and one that is exacerbated by multiple natural and anthropogenic factors. Climate change and a shifting population, for example, highlight the vulnerability and the multitude of stressors placed of water systems. Second, because water availability is intertwined with all aspects of life and development, it has no "stopping point," meaning that the questions enveloping water availability are never quite resolved nor is there one best approach for impacted parties. Third, there is no one-size-fits-all to sustainability addressing droughts—meaning that both the droughts and solutions to have important contextual differences. Fourth, because of the myriad of distinguishable features—there is often not a clear normative answer. Rather solutions must reflect those unique geographical, historical, and contemporary factors as well as values that govern the jurisdiction. Thus, the wickedness of drought policy is a direct result of the uncertainty of risks, complexity of possible solutions, and value divergence of actors (Head 2022). Since there is no single solution to wicked problems, investigating the interactions of multiple variables is required to understand the variety of policy outcomes (Ostrom, Janssen, and Anderies 2007).

Water Policy in the United States

Today's environmental policy (of which water policy is a part of) is built on a series of historical and more contemporary laws and institutions intended to protect air and water quality, manage public lands, protect wildlife, and balance such goals with other pressing priorities such as economic growth and resource development (Klyza and Sousa 2013, 1). In response to manmade and naturally occurring challenges, governments have adopted a variety of approaches and measures depending upon their unique problems and opportunities (Hempel 2009). One subfield of environmental policy is water policy, which is primarily focused on the protection and sustainment of water resources, from wildlife flows to urban delivery.

The stress between sustaining the environment and providing a stable supply of water has led to much of the conflict in water policy. Some states prefer increased water storage in the form of new and larger reservoirs, while others prefer tighter water restrictions and efficiencies. Drought responses in California, Texas, and Alabama exemplify this variability. In Texas, state policymakers increased water storage projects and imposed significant water restrictions on municipal users through a cooperative and structured process (TWDB 2017). Meanwhile in California, the lack of water led to legal battles between competing camps for surface water, followed by comprehensive groundwater reform due to unsustainable withdrawals (DWR 2019a, 2019b). Conversely, in Alabama, the state has initiated the water planning process multiple times, but disbanded the efforts once the drought conditions subsided. Despite similar situations, these states took vastly different paths to address their water insecurities due to the idiosyncratic values, water distribution, and bureaucratic structures (Andreen 2022).

States are generally responsible for water planning with oversight from the Federal Environmental Protection Agency (EPA) to ensure that states meet federal water rules. They are, however, free to structure their environmental programs as desired. The lack of standardization across states has created a myriad of institutions for water policy and enforcement. In California, water is controlled through a State Water Resources Control Board, commonly called the State Water Board, which operates as a child organization of the California Environmental Protection Agency (CalEPA). In Texas, water is regulated by the Texas Council on Environmental Quality (TCEQ), but research and planning are conducted by the Texas Water Development Board (TWDB). Water in Alabama is primarily seen as an economic tool and is managed by the Office of Water Resources (OWR) under the Alabama Department of Economic and Community Affairs (ADECA). Each of these institutions and their policy decisions are shaped by vastly different scopes of authority, values, resources, and approaches (National Integrated Drought Information System 2023).

Most states divide water into surface and ground which are often treated with different legal precedents and rights. East of the Mississippi River, including Alabama, states generally use *riparian rights*, which means that land that borders a river or stream has the first rights to "reasonable" use of water, although the term "reasonable" is open to many interpretations (Smolen, Mittelstet, and Harjo 2017). Based upon English Common Law, this approach is generally used in areas with plentiful water, where landowners are the primary water owners.

In the Western United States, states generally use *prior appropriation*, which is commonly referred to as "first in time-first in right." Under prior appropriation, water rights are primarily owned by the state, which issues rights to users based on a first come, first served basis (Smolen, Mittelstet, and Harjo 2017). In times of shortage, a senior water right holder will usually receive all the water to which he or she is entitled, before a junior user will get any water (Dowell Lashmet 2018). Texas and California are hybrid systems, but primarily built upon prior appropriation, with caveats such as limited domestic and livestock use, wildlife management, and emergency interventions.

Groundwater is historically less regulated than surface water. In Texas, groundwater is property of the landowner, who has the authority to use all the water they can capture, regardless of the effect on neighbors (Texas A&M University [TAMU] 2014). Texas courts have repeatedly denied "reasonable use," meaning that water is not restricted as long as it is not maliciously taken for the purpose of causing injury to a neighbor or willfully wasted (Dowell Lashmet 2018; TAMU 2014). In California, groundwater had been unregulated until the 2014 passage of the Sustainable Groundwater Management Act (DWR 2019a). As currently written though, the act does not fully take effect until at least 2034, leaving groundwater monitored but currently unregulated. Finally, in Alabama groundwater precedent is largely based upon mining operations, holding owners not liable for incidental damages, except for subsidence damages to neighboring wells (Andreen 2022). There are exceptions for "reasonable use" of groundwater in these states, but these cases are rare.

Drought

The National Oceanic and Atmospheric Agency defines droughts along two lines, conceptual and operational. Conceptual definitions are focused on large-scale measures, with terms such as "exceptional" or "extreme" drought conditions and are most often used in water policy. Operational definitions are used to define measures, such as specific deviation from average rainfall and soil evapotranspiration. These definitions are more precise but are often arbitrary, based upon desired measurements and application (National Center for Environmental Information [NCEI] 2023). More specifically, NCEI operationalizes four specific measurements: meteorological, agricultural, hydrological, and socioeconomic.

- Meteorological means that the measured precipitation is a departure from the normal; however, based on climate norms, one state's threshold of drought could be a record rainfall for a more arid region.
- Agricultural means that the moisture in the soil does not meet specific crop's needs. This measure is more difficult in regions that grow multiple crops with different moisture requirements but is generally used in large-scale crop failures.
- Hydrological drought means, or is used, when surface and groundwater supplies are below normal. This is commonly measured in river, aquifer, or reservoir levels compared to historical norm. It is important to note that this is not always tied to precipitation but could be induced if other riparian uses take more water, such as Ethiopia filling Grand Ethiopian Renaissance Dam, which caused a drought in the Nile River valley of Egypt. Similarly, Georgia, Alabama, and Florida have been embroiled in a decades-long dispute over riparian water rights along their border regions due to water withdrawals from growing cities around the Chattahoochee River.
- Socioeconomic means that water shortages begin to affect populations directly. These are the most visible of drought conditions, since a drought

in an unpopulated region is a point of interest, while a drought in a major city is a disaster (Birkland 1997). When populations are included, the term used in water policy often changes from drought to water scarcity, quantifying the water available per individual in the affected region (Rijsberman 2006). In this measure, large population centers are more prone to water scarcity as their per capita water requirement is greater than that in more rural areas.

Drought Severity

The U.S. Drought Monitor releases weekly maps detailing the state of droughts throughout the United States and its territories. The Drought Monitor is a project comprised of the National Drought Mitigation Center, the U.S. Department of Agriculture (USDA), and the National Oceanic and Atmospheric Administration (NOAA). It is measured on a scale of D0–D4, where D0 is "abnormally dry," D1 is "moderate drought," D2 is "severe drought," D3 is "extreme drought," and D4 is "exceptional drought" conditions. What separates the Drought Monitor from other measures is the comprehensive approach to defining drought, combining the inputs from multiple operational measurements into a conceptual analysis. In other words, instead of a purely statistical model of precipitation and water availability, the Drought Monitor combines multiple inputs, including the common Palmer Drought Severity Index, the Standardized Precipitation Index, surface water levels, and snowpack levels, as well as feedback from observers and experts. The multiple sources of information allow a thorough analysis across the varied environments and regions of the United States. In doing so, the U.S. Drought Monitor is one of the most comprehensive and responsive drought tools available (National Weather Service 2023). For this book, the term "drought" will be delineated by the weekly archive of U.S. Drought Monitor's D0–D4 drought scale.

Policy Innovation

The term "innovation" is ubiquitous in every field, from technology to policy, and as a result, the term has innumerable meanings and interpretations. Within relevant policy research, innovation includes "proposed and enacted legislation and regulations as evidence of policy change, or movement in the direction of policy change" (Birkland 2006, 25). Here, we treat policy innovation as the creation of a new program, policy, or management process for the individuals or institutions involved (Steelman 2010, 3). This definition provides two primary advantages. First, this is the definition adopted by Steelman's framework, enabling us to test and subsequently build upon Steelman's original model. Second, this definition encompasses both prior concepts, from policy outcomes to process improvements, thus acknowledging that innovation takes many different forms.

We should note that innovation may be difficult to discern, especially when innovation relates to institutional learning or when the "change" may not result in immediate outputs/outcomes (Birkland 2006). As an example, California's Sustainable Groundwater Management Act (DWR 2019a), a three-bill drought response package, was introduced in February 2014, at a time of only moderate drought conditions in the state, following years of no drought at all (DWR 2019a). However, after the bill was introduced, the state went through a cataclysmic drought, which likely opened the window of opportunity for the bill's passage. The bills were already introduced; however, the subsequent drought led to the adoption of the innovation. To make the problem worse, the bills do not take effect for at least 20 years, so while innovation took place, the innovation has yet to be fully implemented or evaluated.

Another challenge in comparing innovation is that state water institutions serve multiple and, in some cases, disparate purposes. In other words, an innovative approach to effectuate change in one policy arena may have anticipated or unanticipated impacts in another. Similarly, water is viewed differently by each state, shaped by historical developments and varying environmental priorities. Moreover, each state's budget priorities drive much of its water planning; so, while each state may have an environmental management body, funding and the regulatory reach of applicable agencies are likely to differ. Lawmakers in California and Texas, for example, have created powerful institutions with authority to engender sizable policy innovation; by comparison, there remains a much less robust institutional apparatus in Alabama. As such, comparing state institutions that have different goals with varying degrees of authority presents challenges for linking innovation to desired policy outputs and outcomes. Finally, comparing dissimilar water institutions in various stages of development presents case selection limitations. From a development point of view, California's groundwater management reform in 2014 may be more similar to Texas's development in 2002. Similarly, Alabama is still in the infancy of its water policy development. These institutions are not apples to apples, and their respective development is on vastly different stages and paces.

Conclusion

This book is divided into seven chapters. Chapter 1 introduces the audience to the current and interconnected set of challenges confronting state water managers as well as offered a cursory background on droughts, water law, and the states involved. Chapter 2 reviews applicable literature, specifically discussing the development and application of Toddi Steelman's implementation framework in use, policy innovations, planned adaptation, and the role of states in environmental policy. Chapters 3–5 address the three selected cases: Texas, California, and Alabama. For each, there is relevant background information, interview responses, and analysis of drought innovations. Chapter 6 compares across three cases. Finally, Chapter 7 summarizes the findings, notes where this book fits within the larger environmental governance literature, and offers suggestions for future research.

A common (but perhaps apocryphal) phrase that applies to this work is "whiskey is for drinking, water is for fighting over." Although often attributed to Mark Twain, this quote has become more and more insightful as water resources have strained under growing populations, climate change, pollution, and water scarcity. The lack of a comparative analysis of policy innovation following a drought event inhibits the ability to compare unique policy responses and leaves a sizable chasm for understanding the politics of drought management. As such, this volume contributes to the body of work on water policy and provides insight as to why certain states innovate from drought events, while others do not.

References

Alabama Forestry Commission (AFC). 2016. "Drought Emergency 'No Burn' to Remain in Effect Statewide Until Further Notice." December 2, 2016. https://www.rocketcitynow.com/article/news/drought-emergency-no-burn-to-remain-in-effect-statewide-until-further-notice/525-2e92951f-c69e-4f93-b8cc-6b0e9c5f0398

Andreen, William L., and Alabama Water Law (2022). 4 Waters and Water Rights AL-1-AL-52 (Amy Kelley ed., LexisNexis/Matthew Bender 2022). University of Alabama Legal Studies Research Paper No. 4050035. https://ssrn.com/abstract=4050035 or http://dx.doi.org/10.2139/ssrn.4050035.

Birkland, Thomas A. 1997. *After Disaster: Agenda Setting, Public Policy, and Focusing Events*. Washington, DC: Georgetown University Press.

Birkland, Thomas A. 2006. *Lessons of Disaster: Policy Change After Catastrophic Events*. Washington, DC: Georgetown University Press.

Bland, Alastair. 2015. *For California Salmon, Drought and Warm Water Mean Trouble*. January 5. https://e360.yale.edu/features/for_california_salmon_drought_and_warm_water_mean_trouble.

Dingfelder, Jacqueline. 2017. "Wicked Water Problems: Can Network Governance Deliver? Integrated Water Management Case Studies From New Zealand and Oregon, USA." *PDXSchola*. June 1. https://pdxscholar.library.pdx.edu/open_access_etds/3623/.

Division of Water Resources (California) [DWR]. 2019a. "Draft Environmental Impact Report for Long-Term Operation of the California State Water Project." *Department of Water Resources*. November 21. https://water.ca.gov/-/media/DWR-Website/Web-Pages/Programs/State-Water-Project/Files/Deliv-42DEIRv1-112119-Volume-I_ay_19.pdf.

Division of Water Resources (California) [DWR]. 2019b. "SGMA Groundwater Management." *Department of Water Resources*. https://water.ca.gov/Programs/Groundwater-Management/SGMA-Groundwater-Management.

Dowell Lashmet, Tiffany. 2018. "Basics of Texas Water Law." *Texas A&M AgriLife Extension.* January. https://agrilife.org/texasaglaw/files/2018/01/Basics-of-Texas-Water-Law.pdf.

Friedlander, Blaine. 2018. "Groundwater Loss Prompts More California Land Sinking." August 29. https://news.cornell.edu/stories/2018/08/groundwater-loss-prompts-more-california-land-sinking.

Head, Brian W. 2022. *Wicked Problems in Public Policy.* New York: Springer International Publishing.

Hempel, Lamont C. 2009. "Conceptual and Analytical Challenges in Building Sustainable Communities." In *Toward Sustainable Communities: Transition and Transformations in Environmental Policies,* edited by Daniel A. Mazmanian and Michael E. Kraft, 33–62. Cambridge, MA: The MIT Press.

Kingdon, John. A. 1995. *Agendas, Alternatives and Public Policies.* London: Harper Collins. Accessed December 10, 2017.

Klyza, Christopher McGrory, and David J. Sousa. 2013. *American Environmental Policy: Beyond Gridlock.* Cambridge, MA: MIT Press.

Lund, Jay, Josue Medellin-Azuara, John Durand, and Kathleen Stone. 2018. "Lessons From California's 2012–2016 Drought." *Journal of Water Resources Planning and Management* 144 (10): 04018067.

Meinzen-Dick, Ruth. 2007. "Beyond Panaceas in Water Institutions." *Proceedings of the National Academy of Sciences* 104 (39): 15200–15205. https://doi.org/10.1073/pnas.0702296104.

Mekonnen, Mesfin, and Arjen Hoekstra. 2016. *Four Billion People Face Severe Water Scarcity.* Washington, DC: American Association for the Advancement of Science.

National Center for Environmental Information [NCEI]. 2023. "Definition of Drought." www.ncei.noaa.gov/access/monitoring/dyk/drought-definition#:~:text=As%20a%20result%2C%20the%20climatological,%2C%20and%204)%20socioeconomic%20drought.

National Integrated Drought Information System. 2023. "Partners." www.drought.gov/about/partners.

National Weather Service. 2023. "Drought Information." www.weather.gov/mhx/Drought#:~:text=Coming%20out%20of%20drought%2C%20some,Severe%20Drought%20(D2)%3A.

Ostrom, Elinor, Marco A. Janssen, and John M. Anderies. 2007. "Going Beyond Panaceas." *Proceedings of the National Academy of Sciences* 104 (39): 15176–15178. https://doi.org/10.1073/pnas.0701886104.

Persons, Bonnie. 2017. *Water Shortage and Water Law: The Impending Crisis in Semi-Arid Climates.* Accessed October 13, 2019. https://readingroom.law.gsu.edu/cgi/viewcontent.cgi?article=1026&context=jculp

Rijsberman, Frank R. 2006. "Water Scarcity: Fact or Fiction?." *Agricultural Water Management* 80 (1–3): 5–22.

Smolen, Michael D., Aaron Mittelstet, and Bekki Harjo. 2017. "Whose Water Is It Anyway? Comparing the Water Rights Frameworks of Arkansas, Oklahoma, Texas, New Mexico, Georgia, Alabama, and Florida." *Oklahoma State University.* April. Accessed March 2, 2022. https://extension.okstate.edu/fact-sheets/whose-water-is-it-anyway.html.

Steelman, Toddi A. 2010. *Implementing Innovation: Fostering Enduring Change in Environmental and Natural Resource Governance*. Washington, DC: Georgetown University Press.

Texas A&M University [TAMU]. 2014. *Texas Water Law*. https://texaswater. tamu.edu/water-law.

TWDB. 2012. *2012 State Water Plan*. Austin: Texas Water Development Board.

TWDB. 2017. *2017 State Water Plan*. Austin: Texas Water Development Board.

TWDB 2019. *Special Reports to the Texas Legislature*. twdb.texas.gov/ publications/reports/special_legislative_reports/index.asp.

2   A Theory of State Drought Policy

This chapter is broken into four major segments. The first explains the components of the policy innovation and implementation framework with their appropriate literature backgrounds. The second examines the history of water policy with emphasis on collective action and the role of the federal and state governments as policy drivers. The third segment lays the foundation for policy innovation to include disaster response, planned adaptation, and conflict in policy fields. The fourth segment addresses gaps in the literature.

The Policy Innovation and Implementation Framework

Steelman's 2010 policy innovation and implementation framework combines multiple theories and frameworks, including rational choice institutionalism with its focus on the role of individuals in innovation (Geddes 2003), historical institutionalism to help describe the characteristics and importance of structures and past decisions (Thelen 1999; Pierson and Skocpol 2002), and sociological institutionalism that focuses on the larger cultural and cognitive contributions (March and Olsen 1989). The combination of these bodies of literature creates one comprehensive framework that can account for the wide variation seen in policy innovation across culture, structures, and individuals (see Figure 2.1).

As noted, Steelman's framework analyzes policy innovation on three levels—individuals, structures, and culture—in order to show how individuals interact with larger organizational cultures and structures to innovate. A focus on the individual is the foundation of rational choice literature. Geddes (2003, 177) notes: "[R]ational choice arguments use the individual, or some analogue of the individual, as the unit of analysis." Steelman uses the defining features of rationalism throughout her model, including identification of the individual, the individual's goals and preferences, the institutions that constrain options to the individual, and deductive logic (Geddes 2003, 191). For individuals, policy innovation consists of *motivation*: the individuals unsatisfied with the status quo who devise alternative solutions; *norms and*

DOI: 10.4324/9781003498537-2

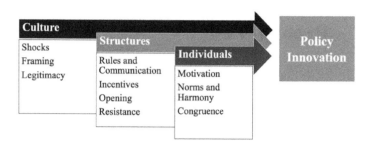

Figure 2.1 Steelman's Policy Innovation and Implementation Framework.

harmony: social norms meant to preserve good working relationships within the organization; and *congruence:* matching of dominant values between the individual and the larger organization.

For structures, Steelman primarily draws upon historical institutionalism and top-down implementation theory to analyze how individuals are encouraged or constrained by an organization's structure and how best to use that structure in order to create innovative policies (Thelen 1999). An individual's use of structures is analyzed through *rules and communication:* organizational compliance towards rules, communication, and information exchange; *incentives:* offsets toward cost-benefit of supporting innovation; *opening:* opportunity for marginalized groups to introduce change; and *resistance:* institutional inertia against innovation.

Finally, for culture, Steelman draws upon sociological institutionalism to help explain how culture organizes meanings by highlighting how individuals identify their place in the larger culture, make sense of their actions and motives, and analyze the identities and motives of others (Ross 2009, 134). Cultural influences include *shocks:* shocks to the system that open an innovation window; *framing:* how problems are presented and perceived by individuals within the culture; and *legitimacy:* how social legitimacy influences which policies a culture may consider (Steelman 2010, 17).

Due to the state level of analysis across a set period of time, our research focuses primarily on the structural and cultural levels. Steelman's *Individuals* variables have previously been operationalized as a particular individual who acted as the primary policy entrepreneur for a specific innovation (Steelman 2010, 179). While individuals are certainly important and will move in and out of this analysis, there are no specific policy entrepreneurs that are the attention of this study. Instead, the focus will be on the cultural aspects, institutional constraints, and environmental characteristics of each state and their potential roles, if any, on the resultant policy innovations. As such, the *Individuals* variables will not be a point of focus in this research.

Characteristics of Culture

Steelman (2010) uses sociological institutionalism to help explain how culture organizes meanings by highlighting an individual's place in the larger culture, how they make sense of their actions and motives, and how they analyze the identities and motives of others (Ross 2009, 134). Cultural influences include shocks, framing, and legitimacy, where shocks describe the actual events; framing describes how the problem is presented to the larger culture; and legitimacy describes how actions and inactions of the organization are perceived by the populace.

Shocks are trigger events that open an innovation window, usually through a crisis that requires policy innovation (Baumgartner and Jones 1993; Kingdon 1995; Birkland 1997). The shocks variable is used as an umbrella to include concepts like focusing events, external shocks, and windows of opportunity (Steelman 2010, 19) and is described in terms of severity, scope, and suddenness. Severity is measured on a scale of impact and institutional response, to include routine event, crisis, disaster, and catastrophe (Faulkner 2001). Scope is largely measured in its effect on humans. As an example, earthquakes, tsunamis, and forest fires can be detrimental in their environmental impact, but they become a disaster when the event impacts humans, specifically the loss of life (Birkland 2006). Lastly, events that occur slowly are less likely to inspire policy innovations, whereas events that are sudden open a policy window, where policy entrepreneurs and interest groups provide potential solutions in the wake of policy failures (Birkland 2006; Kingdon 1995; Baumgartner and Jones 1993).

Framing describes how problems are presented and perceived by individuals within the larger culture. Like Churchill's old adage, *never let a good crisis go to waste*, astute policy entrepreneurs have an ability to frame evidence in ways that suggest a crisis is at hand when they have a potential solution (Mintrom 1997; Stone 1997). Since most policy subsystems enter long periods of stasis, as those systems experience failures, successful entrepreneurs find a way to highlight those failures to policymakers (Baumgartner and Jones 1993). Problem entrepreneurs are thus able to identify, or perhaps exaggerate, the severity of a problem to open a window of opportunity.

A common venue for policy entrepreneurs' attention is the media. In addition to the role of media in exposing the populace to the existence of the event (Birkland 2006), recent research has also focused on the role of media in event selection (Alimi and Maney 2018). The media not only selects which issues and events are reported on, but also provides perceptions and interpretations for the events. Furthermore, the use of opinion pieces allows entrepreneurs to reach a broader audience in their attempts to frame a problem in search of a policy innovation (Alimi and Maney 2018).

Finally, legitimacy explains how organizations seek policy innovations to show social legitimacy within the larger culture (Steelman 2010, 17). Since much of the stasis in government policy is due to risk aversion (Weaver 1986), when potential policy innovations have high visibility and high scope (meaning a severe impact to a large portion of the population), policymakers are hesitant to act due to fear of reprisal in the case of policy failure (Howlett 2014).

In these cases, savvy policy entrepreneurs and interest groups can create environments in which legislators appear out of touch with the electorate if they do not act (Mintrom 1997). Here the risk is shifted from the fear of reprisal for action to the fear of reprisal for inaction. One of the primary tools is the construction of a narrative that occurs through illustrations of settings, characters, plots, and morals (Shanahan et al. 2013). A unique characteristic of a narrative is the use of "devil shift," where authors describe an opponent as evil, while their own side is the "angel" of the story (Shanahan et al. 2013). When actors construct a narrative, from vaccination status to water rights, characters of the story are often depicted as heroes and villains, with a life-and-death plot line and morals tied to the outcome of the story (Shanahan et al. 2013). Often the result of strong narratives spun by policy entrepreneurs, cultural legitimacy is a powerful motivator for environmental organizations.

Characteristics of Structures

For structures, the framework primarily draws upon historical institutionalism and top-down implementation theory to analyze how individuals are encouraged or constrained by an organization's bureaucratic structure and how they may use that structure to create innovative policies (Thelen 1999). Structure is typically analyzed using the organization's rules and communication, incentives, openings, and resistance (March and Olsen 1989, 23).

Rules "define relationships among roles in terms of what an incumbent of one role owes to incumbents of other roles" in order to create predictable behavior within the organization, most often through formalization and structured information (March and Olsen 1989, 23). While formalization creates predictable behavior in the organization, it also limits an individual's actions to their defined roles and communication channels unless there is clear and administrative support for innovative practices (Steelman 2010, 18).

Formalization in an organization can be thought of in two primary forms, schema and scripts (Gioia and Sims Jr. 1986). Schema are mental structures that individuals use to organize knowledge in a categorical fashion, whereas scripts are the series of expected behaviors for a given problem. The use of schemas and scripts creates a structure of expectant behavior within an organization, often in the form of a standard operating procedure.

Structured communication allows for sound decision-making through clear and appropriate information (Steelman 2010). Sense-making across a policy is the key to successful implementation, but differences in interpretation and basic understandings can derail the best policy design (Spillane, Reiser, and Reimer 2002). Formalized communication is a solution to the problem of sense-making by providing structure and rules. Without this, diverse experiences, values, and goals can lead to vastly different policy implementation strategies (McLaughlin 1987) by these "street-level bureaucrats" who are implementing the policy (Weatherley and Lipsky 1977). In many examples, implementers may be creating schema as they go, without the prior experience and frameworks to rely upon (Weick 1979). In the midst of this complexity, successful policy implementation relies upon clear communication about the intent of the policy, the scope of the policy, and the best method of implementation, while also providing a channel for policy innovation when the prescriptive policy is not sufficient (Spillane, Reiser and Reimer 2002). With clear lines of communication, written rules, and unambiguous information, consistent implementation is far more likely (Steelman 2010).

Incentives are enticements the organization provides to encourage innovation, such as grants, funding, or promoting innovative behavior. Incentives are largely used to overcome the "free rider" problem of collective action (Olson 1965). Command-and-control regulation made great strides in environmental adherence but did not result in the transformation of environmental innovation (Press and Mazmanian 2013, 234). On the other hand, pure market-based approaches have also shown to be insufficient for meaningful environmental action. The most successful approach has been market-based incentives that promote self-regulation through financial incentives with the threat of government enforcement (244). From tax breaks to grants, empirical research has shown that institutions that utilize incentives create a greater environment for innovation (Hopkins 2016).

Opening describes the opportunity for marginalized groups to introduce change by encouraging minority opinions and suggestions. Throughout institutional development literature, the importance of outside voices in institutional rules is recurring. From appropriator feedback (Ostrom 1990) to the relative instability of political systems that suppress differing opinions (Chan and Zhao 2016), input from marginalized or minority groups is a sign of a healthy organization. Providing an opening for minority groups allows institutions to create a positive adaptation to the environment (Tyler and Moench 2012), without leading to the type of stasis and results in a large-scale punctuation (Baumgartner and Jones 1993; Chan and Zhao 2016). Opening provides a hospitable environment for incremental change, providing that coalitions compromise by allowing outside voices that may otherwise be marginalized in the policy process (Steelman 2010).

Finally, resistance is the inherent inertia within an organization opposing innovation (Steelman 2010). Just as opening is often the agent of change within an organization, resistance is the agent of stability (Steelman 2010). Constraint on individual behavior is the hallmark of an institution, but these constraints can also induce friction to the point of institutional stasis. To tackle complex problems, a certain amount of hierarchical organization is required (Jaques 1990, 231), but the problem for most organizations is balancing the stability of hierarchical structure with gridlock of bureaucratic processes. Accordingly, resistance to incremental change can leave an organization vulnerable to the large punctuations brought on by shocks that result from the lack of incremental progress (Steelman 2010). These large punctuations, or non-incremental changes, are generally disruptive to an organization as it requires a complete overhaul of rules and scripts, often requiring refinement once a policy is enacted and vulnerabilities are identified.

Water Policy

Common-use resources, such as water, are historically contentious and difficult to regulate. This segment will describe the history of water policy through the lens of collective action, the role of the federal government, and the role of the state governments in environmental policy.

Three primary eras of environmental policy are recognized in the United States, the pre-World War II era, the "golden era" of federal environmental from 1964 to 1980, post-1980 era of state-centric policy (Klyza and Sousa 2013). With the withdrawal of the federal government in environmental policies, state and local governments have become the primary venue for activism (Mazmanian and Kraft 2009, 202; Klyza and Sousa 2013, 229; Bestill and Rabe 2009, 202). Due to this decentralization, local environmental and political idiosyncrasies have substantial impacts on policy outcomes, making policy diffusion between states especially difficult (Mullin 2009).

Collective Action

Scholars have warned of the danger of common-pool resources and the inability of collective cooperation for hundreds of years. In 1832, William Foster Lloyd presented a series of lectures at Oxford University, warning of the dangers of population growth and common resources. Olson (1965) further explained how individuals will seek to "free ride" when groups are large enough to avoid direct oversight. Up to this point, much of organizational theory thought individuals were predisposed to form organizations and herd together as a modern manifestation of tribal tendencies, but Olson provided the example that no state has been able to rely upon mere patriotism or goodwill to support itself; instead, states must use compulsory taxes to ensure participation or other noncollective goods to provide incentive (13–16). However, when groups are small enough to ensure personal connection and supervision, individuals are more likely to contribute to the collective good (54).

Hardin (1998) updated these ideas when in the "tragedy of the commons," a dystopian future where the combination of expanding populations and limited common-pool resources would be locked into a spiral toward destruction due to self-interested overuse. This theory of profit maximization assumed that users would be unable to establish controls over resources leading to the exhaustion of the environment.

While some scholars predicted an overpopulated world with endless competition for resources, others held on to optimism in the capacity of institutions as agents of cooperation (King, Keohane, and Verba 1994). History has shown a rich tradition of cooperation over environmental common areas (Cox 1985), and although resources are certainly stretched, the capacity for ingenuity that has enabled human cooperation up to this point remains (Homer-Dixon 2002). Previous examples of the prisoner's dilemma had a fatal flaw when they assumed that the games were finite and that actors would settle for an immediate benefit over a long-term reciprocity (Ostrom 1990, 2008). Iterated prisoner's dilemma models showed the ability of actors to create institutions as a means of providing information, reducing transaction costs, and altering the payoff associated with cooperation (Axelrod 1984). The combination of these "tit-for-tat" relationships, along with modern technological advances, provided a hope for future cooperation (Homer-Dixon 2002).

To add to this optimism, Ostrom (1990, 2002, 2008) noticed that despite the challenges to collective action, individuals have repeatedly shown the ability to self-organize and create institutions for resource management. The simple answer appears to be the ability for actors to communicate. With open communication, actors can establish institutions for cooperation (Axelrod 1984); however, for these institutions to be successful, certain principles are required (Ostrom 1990). For Ostrom (1990), the answer is not a one-size-fits-all, but rather a customizable solution of public and private exchanges as the market and the environment dictate. While the exact composition of institutions varies with the environment, certain trends repeatedly appear. Successful institutions tend to have characteristics such as clearly defined boundaries of use, collective choice rules, behavior monitoring, conflict resolution mechanisms, and the ability to self-organize apart from government agencies.

Research into self-organizing farming groups shows that the rules in place are unique to the environment with strong self-regulation and cost sharing (Regmi 2008; Lam and Chiu 2016). With external oversight, officials are not as motivated to ensure the equitable access and less likely to enforce violations. Government agencies frequently overemphasize the physical structures of irrigation systems, versus the relational side of institutions (Ostrom 1990, 2008). In the name of efficiency, often outsiders are brought in, large projects are praised, and the years' worth of goodwill and cooperation are discounted in the name of progress (Regmi 2008). The result is often far worse outcomes for the end users.

Government intervention appears to be most effective when adding resources to an existing institutional arrangement (Lam and Chiu 2016). Bureaucratic oversight from a remote location cannot keep up with fragile agreements in a changing environment. Lower-level control is often required, with the support of government resources (Ostrom 1990). While Olson (1965) assumed that group size directly correlated to outcomes, Ostrom (1990) has shown that the relationship is more complex. Large collectives of locally run irrigation projects can be successful, despite the overall size (Meinzen-Dick 2007). In this sense, it is not the size of the collective that predicts success or failure, but the mixture of state, collective, and market institutions.

This is not to say that participants will always self-organize successfully. On these lower levels, characteristics such as strong leadership, previous history of cooperation, and a shared understanding of the necessity to organize tended to be associated with positive outcomes (Ostrom 1990). The importance here is that in common resource systems, the relationships within the institutions are often a far better predictor to success than the physical structures (Meinzen-Dick 2007). Successful institutions start with relationships between users, not efficiency of resources.

The Role of the Federal Government in Environmental Policy

Traditionally, the federal government's role in environmental policy is limited, leaving most management to the states. The one exception to this rule was a result of the Supremacy Clause in the U.S. Constitution, where the federal government managed interstate trade through infrastructure such as roads and waterways (Abrams 2009). The Army Corps of Engineers was the primary actor in this role, which led to nationwide improvements in locks and dams, dredging, and navigation equipment.

While the federal government largely stayed within the self-imposed constraints, there were a few exceptions. One of these is the Central Valley Flood Protection Plan (CVPP) project, which is a consortium of locks and dams that funnel water from water-rich northern California to the arid but fertile Central Valley. As opposed to interstate travel, the CVPP was wholly located in the state of California and used primarily for irrigation, not navigation. While this project will be covered in detail later, it is important to note that this was originally a state project, as part of the California Central Valley Project Act in 1933; however, the Great Depression left California unable to pay for the expense (CVPP 2022). The federal government saw an opportunity to alleviate some pressure from the Dust Bowl droughts in the Midwest and allocated funds as part of the recurring Rivers and Harbors Act in 1935, but this foray into irrigation and state infrastructure was limited in its scope. Following World War II, Washington took the lead with large-scale infrastructure, such as interstates and universities, but largely continued the precedent of leaving land and water management to the states (Barbour and Teitz 2009).

The Water Supply Act of 1958 (43 U.S. Code § 390b) was the first major change to this policy and greatly expanded the federal government's role in water policy. Although the states were still primarily responsible for water management, the federal government began to take an active role in the federal navigation, flood control, and irrigation projects, establishing a hierarchical federal precedence over state enforcement, particularly in terms of water quality monitoring and enforcement. The national government had previously passed the 1948 Federal Water Pollution Control Act as a way to prompt state action, but the bill had little in the way of enforcement (Morris 2022; Vig and Kraft 2012). With little action from the states and public outcry over extreme pollution, such as California's air quality and Ohio's infamous fires on the Cuyahoga River, federal action was required. The "golden era" of federal environmental lawmaking, between 1964 and 1980, saw the U.S. Congress pass 22 major laws addressing environmental management and pollution control, most of them were overwhelmingly bipartisan (Klyza and Sousa 2013). Included in this legislation was Clean Air Act and Clean Water Act, as well as the establishment of the Environmental Protection Agency (EPA). This period was epitomized by the strong centralization and command-and-control nature of federal regulation and the national elevation of environmental concerns (Mazmanian and Kraft 2009).

As deregulation took hold in the 1980s, largely the result of conservative reaction to federal regulation and push for government efficiency, environmental policies were lowered back down to the state and local levels (Mazmanian and Kraft 2009). While subsequent administrations have pushed various environmental policies, funding for environmental policy has been in slow but steady decline since the early 1980s (Vig and Kraft 2012, 18).

Despite this decline, federal funding is still a powerful force in state policies. Federally funded policies diffuse through states about twice as fast as state-only funded policies, especially if those policies are positively incentivized through rewards, vice negatively incentivized through punishment (Welch and Thompson 1980). Furthermore, states have inherent pressure from federal mandates or recommendations that have become normative, especially with the threat of future federal pressure (Berry and Berry 2018).

Since the early 1980s, gridlock has become the norm for federal environmental policy. The federal government has moved from lead actor to supporting roles as regulator and financier. This has left the vacuum of power in policy, creating a unique opportunity for states and activist groups to take the lead in policy innovation.

The Role of States in Environmental Policy

Since the end of federal leadership during the "golden era" of environmental policy (1964–1980), state and local governments have been instrumental in environmental activism (Bestill and Rabe 2009, 202; Klyza and Sousa 2013, 229). From cities embracing the Kyoto Protocol to states managing endangered species designations, elaborate networks and partnerships have formed between like-minded states, resulting in successful policy diffusion across borders, with little policy oversight from Washington (Bestill and Rabe 2009, 202). As of 2022, 47 of the 50 states had completed inventories of greenhouse gasses, while 14 states had accepted California's stringent car emissions standards (EPA 2022; Mazmanian and Kraft 2009, 205). These pacts have resulted in rather large intrastate networks and interstate policy diffusion, from regionally based laws to joint lawsuits.

The true weight of subnational governments became apparent in the U.S. Supreme Court's 2007 decision in *Massachusetts v. EPA*, where the court sided with Massachusetts over the federal resistance to label carbon dioxide as an air pollutant under the Clean Air Act Amendments. Since then, successive states and municipalities have sought the legal option to force federal environmental action, while also providing a venue for activist state attorneys general to pursue state climate plans (Bestill and Rabe 2009).

There are two primary forms of policy innovation for states, internal determinants, and diffusion models (Berry and Berry 1990). Internal determinants are largely the result of the characteristics and needs of the state, whereas diffusion models are when states look to emulate successful policies from other states (Mintrom 1997). While early research sought to separate these two mechanisms in "single explanation" methodologies, the policy innovation process is often a mix of these, where states look to find other successful external solutions to their internal problems (Berry and Berry 2018). Furthermore, state officials can feel pressure from like-minded states as a result of shared norms and expectations of ideological viewpoints (Berry and Berry 2018). The decision to pursue environmental policies is a complex one at the state level, well beyond simple answers. The interaction between environmental, financial, and cultural factors creates a complex policy field; a true policy laboratory (Travis et al. 2004).

States have multiple venues for citizen engagement in environmental policy, to include traditional legislatures, ballot initiatives, and litigation, which allows for a more nuanced and customizable environmental policy (Klyza and Sousa 2013, 259). It also enables varied policy implementation, allowing a more robust test of policies across multiple environments. Due to the unique nature of distribution and the variety of sources, water has a particularly local flavor (Mullin 2009). A single mountain ridge can separate the water rich from the water poor areas, and riparian laws can create an environment where a farmer that is merely miles away from a major river source has no legal right to irrigation water. The idiosyncratic nature of water policy requires customized and nuanced policy that can only be achieved at lower levels.

Some states are so geographically diverse that planning must be pushed to lower levels of planning in order to achieve granularity, again introducing complexity to the policy (Barbour and Teitz 2009, 175). While opening the "black box" of state and regional planning adds complexity to research, states are the current battleground for environmental policy innovation (Bestill and

Rabe 2009, 222), as such, the state is the appropriate level for comparing water policy in the United States.

With the lack of federal leadership, most states have attempted to secure water resources by creating state water plans. Serious water planning is most often found in the American West, where water has been a commodity since the days of westward expansion. Casado-Perez et al. (2015) provide an excellent review of western water plans. Ranging from 80 pages in Utah, to an over 1,000-page, three-volume set in California, the difference in planning among western states is significant (Casado-Perez et al. 2015).

This planning can also be the source of conflict, especially when multiple states compete for shared water resources, which can ultimately force federal intervention. From battles over the rights within Alabama, Florida, and Georgia (Andreen 2016, 2022), to over the withdrawal of the Colorado river between California and Arizona, one state's water policy directly affects joint riparian neighbors and is an opportunity for cooperation but also a source of conflict. Moreover, we should note that climate change has amplified many of these opportunities and stressors.

Innovation

Innovation in environmental policy is often in response to disasters but is heavily influenced by multiple variables. Successful states must be willing to learn from disasters by investing in new institutions (North 1990) and establishing planned adaptation through continual reviews of standing policies (Vogus and Sutcliffe 2007). Unfortunately, the work spent in investing in innovation can be hampered by an increasing number of actors in a policy subsystem (Birkland 2006) and the lack of a cohesive environmental culture (Besharov 2014).

Learning From Disasters

In the aftermath of a disaster, institutional learning generally occurs; however, the political will must exist to apply the newly acquired information. Building resilience within the system is necessary to provide a positive adjustment under challenging conditions such that the organization emerges from those conditions strengthened and more resourceful (Sutcliffe and Vogus 2003; Vogus and Sutcliffe 2007). For water, anticipation of drought events can lead to multiple responses—often corresponding to environmental factors and political ideologies—to include water use restrictions, water distribution changes, and building slack resources through water storage. Each of these options is idiosyncratic but with the desired end goal to "bend, but not break" in a recurring drought (Vogus and Sutcliffe 2007).

One of the most important outcomes from policy learning is the creation of organizations and institutions that allow for future policymaking.

Organizations are agents with preferences and objectives, while institutions are social constraints that restrict behavior (Khalil 1995, 445). To put it more simply, institutions define the rules of the game, whereas organizations are the players (Scheberle 2004, Scott 1995). Organizations are often created suddenly as a collective response to an acute need, such as the creation of the EPA during the height of federal environmental policy. But from this example, this organization needed to find an identity and purpose, so it created the "institutions," or expectations of organizational and individual behavior gradually over time (Scott 1995, 18).

Organizations take a significant amount of energy and political will to create, but they serve the purpose of preventing the wholesale re-creation of knowledge with every new policy problem (Lindblom 1979). The importance of an organization, such as a water board whose defined purpose is to monitor water levels and recommend policy changes, cannot be overstated. While these organizations can vary in effectiveness based upon institutional characteristics, their very existence in the face of a focusing event allows for the "exchange of opinions, beliefs, and theories about why the event happened and whether existing policy can address the problems revealed by the event" (Birkland 2006, 23).

In summarizing the characteristics of resilient versus weak organizations, the outcome can be described as either a positive adaptation to the environment or a negative adaptation that can lead to organizational failure (Tyler and Moench 2012). Positive adaptations describe institutions that allow maximum flexibility in the face of adversity, while institutions that are slow to react to changes result in instability through inaction.

States that experience frequent catastrophic droughts would be expected to have different policies and institutions than states with water abundance. Since there is no panacea of institutional solutions, each state must choose the appropriate response based upon their myriad of challenges and risks (Ostrom 1990), but these choices will be limited by the existing institutions and political culture. With increasing environmental degradation due to shifting climate patterns and global warming (Tyler and Moench 2012), states must have the organizational capacity to absorb shocks and bounce back from both anticipated and unanticipated dangers (Wildavsky 1988). These adaptations can either be the result of failed policies that lead to institutional learning, or they can be an organized and purposeful adaptation based upon a retrospective analysis.

The goal of environmental policy is disaster mitigation (Birkland 2006, 106). Often, policies are focused on the suffering as the result of disasters, rather than mitigating the disasters to begin with. It is easier to justify spending money based on the images of a community in ruins than on preventing the next emergency. However, in 2000, the federal Disaster Mitigation Act (DMA) required states to create state hazard mitigation plans (Berke, Smith, and Lyles 2012).

Several characteristics of disaster mitigation have shown to lead to successful policies, to include goals and desired conditions; empirical foundation to hazard identification; clear policy guide to future decisions; monitoring of organization responsibilities; coordination between state and local organizations; and inclusion of other than government organizations (Berke, Smith, and Lyles 2012). State mitigation plans vary in their quality; however, much of this variation could be a result of their use, or lack thereof. States that repeatedly face major and varied hazards, from wildfires and mudslides to earthquakes and droughts, such as California, have more refined hazard plans (Birkland 2006). Conversely, states that do not face regular weather or environmental disasters, such as the New England region, have less developed hazard mitigation plans (Berke, Smith, and Lyles 2012). With this understanding, state hazard response is more tied to experience than political affiliation. For this research, policy innovation outcomes will be observed in existing political institutions and mechanisms that allow for iterative environmental planning and feedback, as well as in policy innovations resulting from drought events.

Planned Adaptation in Water Policies

Planned adaptation is the use of internal and external processes to self-correct as conditions change over time (McCray and Oye 2007). From the military to the EPA, anticipating future environments is one of the most difficult tasks that any organization can undertake, as well as expensive over time (McCray and Oye 2007). Since most long-term predictions are incorrect, policies and regulations must be continuously updated and revised as new information becomes available to achieve the desired outcome, *if those original outcomes are still valid.* This continuous adaptation requires recurring, multi-year reviews and the incorporation of new data, with the institutional support.

Three primary factors influence an incorrect prediction by policymakers: changes in the external environment, the internal environment, and organizational capacity (Watts et al. 2007; Ostrom 1990). Since most environmental policy is based on inductive logic (Bruce 2006), as new changes occur, policymakers must change policy to account for new data. Ostrom (1990) noted that the unintended consequences from incorrect assumptions must be accounted for through adaptive processes, otherwise policies will quickly become ineffective. With existential resources such as water, policies much be continuously adapted as the data changes, resulting in divergent policies with varying environmental inputs. States that either have no long-term planning or do not invest in planned adaptation run the risk of falling behind the institutional resilience required for a changing environment. By their nature, water providers cannot wait until a catastrophic failure to implement change (Birkland 2006). This requires a constant reevaluation of their internal capabilities, environmental threats, and slack in the system.

Number of Actors in Fractured Policy Subsystems

One of the factors that can derail cooperation and shared policy learning is infighting between actors in the policy subsystem. Actors are individuals, organizations, or interest groups that interact within the policy process. It is important from an analytical standpoint to highlight that actors are both creating the system and acting within the system (Scott 1995, 18). In this sense, some powerful actors look to impose institutional constraints on other actors—even subsequent governmental actors—to ensure their desired outcome is achieved (King, Keohane, and Verba 1994).

Actors form policy domains within policy subsystems through the joining of shared core policy beliefs, as long as their shared beliefs do not conflict with deep core policy beliefs (Sabatier 1999). One of the advantages of actor-centric research is trying to explain the complexity of intricate webs and relationships, vice the structure of institutional analysis (Heclo 1978). As an example, logic would dictate that the more actors and coalition members behind a particular initiative, the more powerful that movement becomes; however, there comes a point where too many actors can fracture the shared identity that allows for collective action and derail the policy process.

As the number and diversity of actors increase in a coalition, those actors will be pulled apart by their deep core beliefs (Birkland 2006). As a practical example, when a single environmental group is negotiating with multiple industry groups, they are more likely to compromise to achieve some net positive result (Steelman 2010). Conversely, when multiple environmental groups are collectively negotiating, they are more likely to prove their commitment to the larger environmental cause and are less likely to compromise.

Actors adapt their policy inputs through selective attention, subjective assimilation, emotional conditioning, and a disproportionate view of political wins and losses (Weible et al. 2011). Convergences and divergences in each of the variables can lead to cooperation and conflict within policy subsystems, making strange coalitions at times, while also driving apart similar groups over some seemingly small differences. Research has shown that the optimum number of actors in a coalition is two to four (Weible et al. 2011; Birkland 2006). As this number increases, infighting over deep core beliefs causes fracturing among the actors, diminishing their collective power (Birkland 2006). Research into environmental policy forums found that forums with five members had a 59 percent chance of achieving a mutually beneficial outcome, while in forums with 30+ members, the chances of cooperation drop by over 20 percent (Lubell, Mewhirter, and Berardo 2020). Furthermore, if there are opposing policy coalitions, the chances of a successful policy implementation diminish greatly (Birkland 2006).

Researchers have long assumed the inevitability of conflict in policy (Follett 2017), yet the underlying mechanisms have largely gone untraced. Certain variables lead to conflict in policy subsystems, such as the unwillingness of groups to compromise, the perceived political costs of cooperation, the proximity to the specific problem, the perceived morality of the problem, and the difficulty in understanding to complex problems and solutions (Weible and Heikkila 2017). While many of these factors seem logical, providing a framework for understanding conflict has been largely ignored until recently.

A key driver to policy conflict is when a policy position from one group incurs potential costs on the identities and values of another group (McAdams, Tarrow, and Tilly 2001). These costs may not be intended or even originally identified, but as policies take shape, understanding the second- and third-order effects can greatly influence cooperation or conflict for the future. Additionally, a history of conflict adds to the "repertoire" of counteraction between groups (McAdams, Tarrow, and Tilly 2001). In this sense, just as groups can create a record of cooperation over time (Axelrod 1984), they can similarly create a record of conflict, making conflict the default interaction between actors (Morris et al. 2013).

Groups that concentrate on a single issue, such as water quality or CO_2 reduction, are more likely to create conflict in that subsystem and be less likely to compromise on potential solutions as they see the world as zero-sum for their single interest (Lubell, Mewhirter, and Berardo 2020). When opposition groups, such as environmental and farming interest groups, are at the extreme ends of the policy spectrum,, cooperation is unlikely, leading to conflict that can only be resolved by a well-reasoned state agency or by the courts (Hanemann and Dyckman 2009).

Conflicting Cultures Leads to Less Innovation

One of the foundational ideas of organization theory is the concept of *cognitive consensuality*, or how the cognitions of individuals become coalesced into a singular organizational purpose and desired outcome (Gioia and Sims 1986, 8). Without this socially constructed view of the world, organizations cannot aggregate individuals toward a goal, and the more diverse the opinions and views of a group, the less likely that group will achieve its stated purpose.

That said, not all individuals of an organization or a policy subsystem are equally geared toward the organization's goals. Individuals and organizations that identify as "idealist" are less likely to compromise their views and values, as these values are the core of their identity (Besharov 2014, 1502). Idealists tend to view a policy stance as morally correct, as opposed to others who may have a policy preference but are willing to compromise to achieve some progress, even if it is not their fully desired outcome. Idealist individuals and groups are more likely to place their own views and values above the organization's and are thereby willing to prevent organizational innovation if they feel the perceived progress is insufficient (Besharov 2014; Liden et al. 2014).

Idealist groups tend to weigh heavily on the good versus evil narrative (Shanahan et al. 2013) and are often looking for the "big win" that will justify their "righteous indignation" (Steelman 2010, 108). Social problems, such as the environment, are of such magnitude that it can be overwhelming for groups to tackle. Making small changes can collectively have a big effect, but this often does not match the vision of idealist groups. As an example, the EPA's founding director, William Ruckelshaus, focused his first few days on small-scale lawsuits that were clear-cut victories (Weick 1984). This established a foothold of success for the EPA, which helped cement its culture in the early days. Unfortunately, in order to justify their existence, many environmental groups create a narrative of doom and gloom, urging massive reconstructions of the economy. While these characterizations create headlines and interest, they force the organizations to go after large wins from the beginning, instead of smaller compromises that can collectively lead to progress (Weick 1984, 48).

Strong leadership within a policy subsystem can help align values of disparate groups. Compromise is the root of iterative progress, and managers who enable justifiable compromise are far more likely to avoid value conflicts within their organizations (Bryner and Duffy 2012). The assumption of compromising research is that the managers can negotiate and have some ambiguity within policy itself. Policy innovations within homogenous groups that will be equally impacted by the enforcement costs are far more likely than heterogenous groups that will face varying degrees of costs (Scheberle 2004, 46). However, policy innovations among actors with different values are difficult because with multiple potential outcomes, final decisions create an environment of winners and losers (Bryner and Duffy 2012).

If one side has "lost" in the past they are less likely to enter compromise again. This is especially true if the consequences were not overtly stated, but rather realized once the policy was in place (Bryner and Duffy 2012). A history of perceived wrongs creates tension where there may have been goodwill before. After all, prospect theory has shown that individuals remember their losses more than their gains, making them more prone to future conflict than future cooperation (Quattrone and Tversky 1988).

Citizen Impact and a Cohesive Culture

As previously stated, the framing of a disaster becomes much easier when there is a human toll, especially the loss of life (Birkland 2006). The ability to attach a human interest to the story is a powerful narrative that piques citizen engagement and support. It is through these narratives that citizens become engaged in policy solutions and open the windows for implementation (Kingdon 1995; Birkland 1997).

Social constructions are one of the core beliefs of modern sociology (Berger and Luckmann 1966), which is a salient characteristic of citizen engagement as it explains much of the subjectivity of bounded rationality (Simon 1957). To simplify the complexity of decisions, individuals use emotions and heuristics to reduce the number of options to a more manageable selection, which are hidden behind the "normal" processes of the brain (Birkland 1997). It is through these heuristics and emotions that individuals form their construction of reality. With that said, individuals search for existing constructs that closely match their own views, often found in political parties and coalitions, and conform their views in search of acceptance and legitimacy (Berger and Luckmann 1966). Collectively, this process of constructions allows individuals to categorize complex issues into simple political bins and categorize complex individuals into good and evil.

Citizens perceive their reality and the plurality of interpretations through emotionally charged illustrations of settings, characters, plots, and morals (Shanahan et al. 2018). Modern psychological literature illustrates that emotion is not a side effect of decision-making and memory, it is a fundamental requirement (Bechara, Damasio, and Damasio 2000). By overlaying the emotion of good versus evil over a policy, narrators form powerful connections and memories with readers. This heightened state of awareness is often mobilized through imagery and videos, which form compelling, lasting memories (Birkland 2006).

The real question for policy entrepreneurs is when citizen perceptions open a "window of opportunity" and how long that window can stay open (Herweg, Zahariadis, and Zohlnhoffer 2018). Some windows are predictable, such as post elections, while others are not, such as following a disaster. Environmental policy is a mix of both. For states that experience recurring disasters, the windows of opportunity are unpredictable in exact timing, but history says that they will occur and will likely open an opportunity to join the political, problem, and solution streams, if a reasonable solution exists (Kingdon 1995; Birkland 2006; Herweg, Zahariadis and Zohlnhoffer 2018). In these cases, savvy policy entrepreneurs can wait with ready-made solutions for the right opportunity to take advantage of a shocked public. This idea of recurring widows is an interesting twist on the "issue-attention cycle" (see Downs 1972) of public consciousness. The cycle consists of the pre-problem stage, alarmed discovery and euphoric enthusiasm, realization of the costs, gradual decline in intense public interest, and finally the post-problem stage. The cycle is complete when the problem has effectively faded from the public attention (Downs 1972, 41–42).

While droughts are finite in their duration, a disastrous drought can affect almost all citizens simultaneously, and perhaps even change their daily lives. Major droughts, from the ones in this project to the South African drought in Cape Town (Parks et al. 2019), illustrate how citizens' attitudes enable collective cooperation to achieve long-term solutions.

From a public attention perspective, the middle three stages of the cycle are the most important: alarmed discovery and euphoric enthusiasm, realization of the costs, and gradual decline in intense public interest (Petersen 2009). Without the alarmism and euphoria, most citizens will not be willing to bear the cost of intervention. Thankfully, most events will never get to the disaster level; however, this framework has utility for this research project. Downs' entire model is built on the environmental push of the 1970s, where narratives of "killer smog" and small groups of "villains" permeate though the media coverage of the day (Downs 1972, 47). For most citizens, the source and quantity of water is something that they never consider when they turn on the faucet. Until citizens are faced with the hard reality of unreliable water, they will rarely devote their attention to the topic (Petersen 2009).

Voters are perhaps myopic in the sense that they are unwilling to support spending on disaster preparedness before a disaster occurs (Healy and Malhorta 2009). Most federal disaster policy focuses on rebuilding after a disaster, vice future mitigation (Birkland 2006, 110). Once the immediate rebuilding has been completed, the public loses interest in the problem and moves on to other, more pressing matters. Large-scale preparedness is often possible only after a galvanizing focusing event, regardless of the proximity of the event, and ironically, it is sometimes disasters that strike other areas that can lead to change (Healy and Malhorta 2009).

This highlights one of the unique features of environmental policy. As stated, the timing of droughts is unpredictable, but the inevitability in certain regions, such as the American Southwest, is assured. Given the recurrence of droughts, these citizens never truly enter the post-problem stage of the model (Petersen 2009). Yes, the problem of drought fades of intense public scrutiny during water-rich years, but there is always the looming threat of the next one in drought-prone areas. In this sense, the issue-attention cycle never completes, it just stays in a dormant fourth stage until it is catapulted back into the second stage again (Petersen 2009). Furthermore, once an issue has been through the cycle, it is able to quickly regain the public consciousness (Downs 1972).

Successive disasters create an opportunity for policy leaders to make substantive changes in certain policy fields. As an example, in 1953, a disastrous flooding in the Netherlands killed over 1,800 people and helped shape a significant policy shift in mitigating future disasters (Meijerink 2005). While substantive changes occurred, the government's research into other flood control measures was stalled by coalitions of actors who would be negatively affected by further controls. When river floods in the Netherlands in the 1990s caused the evacuation of over 200,000 residents, the existing flood control measures were quickly enacted and changed the Dutch identity from one that reclaimed too much sea area as lowlands to one that sought "living with the water" (Meijerink 2005). The fact that so many citizens were directly impacted by the threat is that the flooding allowed for a cooperative environmental culture, opening the window of opportunity for policy change.

Unfortunately, not all policy debates allow for the coalescence. Natural disaster policies, such as flooding and earthquake mitigation, often do not have interest groups opposing their implementation, only the time and money that the agenda allows (Birkland 2006). Conversely, policies such as water storage and climate change are highly contentious on both sides and largely prevent the cooperation required for a cooperative policy subsystem. Finding common ground is much more difficult in their areas, and the resultant conflict prevents iterative policy improvements.

Gaps in the Literature

Up to this point, Steelman's (2010) framework has mostly been employed in individual case studies. This research will be the first known broad application of the framework as a state comparative analysis, although some research has used single variables in a comparative manner (Davis 2012). Because of the limited applications, the framework has no mechanism for environmental variance between cases. To account for this, this study utilizes grounded theory to identify and implement environmental variables into the research. This contributes to the literature by adding these missing environmental conditions to Steelman's framework with the hope of increasing its explanatory power.

The effect of the local environment on political institutions and policies is undeveloped. Most water research is based on single events or locations, frequently geared around disasters. There is little analysis of intuitional development and interactive process. Furthermore, research into policy conflict (Lubell, Mewhirter, and Berardo 2020) has called for a more qualitative approach to help understand the dynamics of polycentric institutions and comparative analysis (Greif 1998). This book will allow for a state-centered case study, while also providing comparisons across multiple environments. It will provide institutional-level analysis of policies, while explaining some of the variation between idiosyncratic state policies.

A Note on Study Methods

This research employs qualitative analysis to study the interactions between institutions and their environments. We utilize a collective case study of water institutions and policies (Creswell and Poth 2018, 150) with diverse case selection (Seawright and Gerring 2008) of three U.S. states across the full variation of political spectrum and water availability. The three selected states—Alabama, California, and Texas—allow us to compare institutional and organizational interactions with their local environments.

Case Selection

Since most drought research is focused on water scarce areas, such as the American Southwest, there is a gap in knowledge for states that are normally flush with water but still face droughts (King, Keohane, and Verba 1994, 44). Due to its natural water abundance, Alabama represents an extreme case compared to most drought research; however, it provides a more typical case compared to states east of the Mississippi River.

Texas and California are included for two reasons. First, these are the two largest states in the lower 48 by both territorial size and population. Territorially, they are quite different, but each faces natural obstacles and geographic limitations that are ideally suited for comparative analysis. From a population perspective, according to U.S. Census Bureau estimates, one-fifth of the U.S. population lives within these two states, meaning that their state policies are particularly impactful to the overall population.

The second reason for the inclusion of these states is their history of catastrophic droughts, creating the need for extensive water planning through policy innovation and implementation. Each of these states has experienced impactful droughts that have left a legacy of public memory, which exemplifies the extreme of water policy in the united States (Seawright and Gerring 2008, 297). Due to this history, they each have robust water planning processes and policy reviews; however, the resultant policy innovations from these two states vary significantly due to the environmental differences and path dependency. In these cases, interviews with key decision-makers provide insight into the schema and scripts (Gioia and Sims 1986) of these organizations to better understand how their institutions were formed (Yin 2016).

Research Design

We employed the diverse case method (Seawright and Gerring 2008) for our three cases for the years 2011–2016. This time frame was selected as a control for similar drought conditions, policy diffusion, and collective learning. Droughts were measured and dated using the U.S. Drought Monitor. For water board meetings, we analyzed six years' worth of meetings from the various states to assess opening, public participation, and narratives used in framing the problem. In Alabama, meeting minutes and videos from the Alabama Environmental Management Commission were analyzed. In California, the State Water Control Board was analyzed, whereas in Texas, local and regional water boards, to include Dallas, Ft. Worth, and Texas Region C, were analyzed since the state water board meetings were not available during the time of this study.

Video and phone interviews were conducted with key personnel and decision-makers from each of the selected cases. Over 50 key decision-makers were identified for interviews and contacted about participation; however, access across the states varied greatly. California and Texas were in the midst of droughts, which influenced many state-level leaders to decline participation. Meanwhile, Alabama was limited by the small number of participants in the water process. Final participants were assigned random letters for anonymity.

The interviews were primarily focused on understanding five major variables in each state: the water planning process, the frequency and utility of reviewing existing water policy, the importance of shocks in driving policy innovations, cooperation or conflict in the state environmental culture, and the impact of the environmental distribution of water on policy outcomes. The interview protocol for the interviews is presented in the Appendix.

References

Abrams, Robert H. 2009. "Water Federalism and the Army Corps of Engineers' Role in Eastern States Water Allocation." *University of Arkansas at Little Rock Law Review* 31 (3): 395. Accessed October 31, 2019. https://lawrepository.ualr.edu/cgi/viewcontent.cgi?article=1117&context=lawreview.

Alimi, Eitan Y., and Gregory Maney. 2018. "Focusing on Focusing Events: Event Selection, Media Coverage, and the Dynamics of Contentious Meaning-Making." *Sociological Forum* 33 (3): 757–782.

Andreen, William L. 2016. "No Virtue Like Necessity: Dealing With Nonpoint Source Pollution and Environmental Flows in the Face of Climate Change." *Virginia Environmental Law Journal* 34 (2): 255–296.

Andreen, William L., and Alabama Water Law. 2022. 4 Waters and Water Rights AL-1-AL-52 (Amy Kelley ed., LexisNexis/Matthew Bender 2022). *University of Alabama Legal Studies Research Paper No. 4050035.* https://ssrn.com/abstract=4050035 or http://dx.doi.org/10.2139/ssrn.4050035.

Axelrod, Robert 1984. *The Evolution of Cooperation.* New York: Basic Books.

Barbour, Elisa, and Michael B. Teitz. 2009. "Blueprint Planning in California: An Experiment in Regional Planning for Sustainable Development." In *Toward Sustainable Communities*, edited by Daniel A. Mazmanian and Michael E. Kraft, 171–200. Cambridge, MA: MIT Press.

Baumgartner, Frank R., and Bryan D. Jones. 1993. *Agendas and Instability in American Politics.* Chicago: University of Chicago Press.

Bechara, Antoine, Hanna Damasio, and Antonio R. Damasio. 2000. "Emotion, Decision-Making and the Orbitofrontal Cortex." *Cerebral Cortex* 10 (3): 295–307.

Berger, Peter, and Thomas Luckmann. 1966. *The Social Construction of Reality: A Treatise in the Sociology of Knowledge.* New York: Doubleday.

Berke, Philip, Gavin Smith, and Ward Lyles. 2012. "Planning for Resiliency: Evaluation of State Hazard Mitigation Plans Under the Disaster Mitigation Act." *Natural Hazards Review* 13 (2): 139–149.

Berry, Frances Stokes, and William D. Berry. 1990. "State Lottery Adoptions as Policy Innovations: An Event History Analysis." *American Political Science Review* 84 (2): 395–415.

Berry, Frances Stokes, and William D. Berry. 2018. "Innovation and Diffusion Models in Policy Research." In *Theories of the Policy Process*, 4th ed., edited by Christopher Weible and Paul A. Sabatier, 253–297. New York: Routledge.

Besharov, Marya L. 2014. "The Relational Ecology of Identification: How Organizational Identification Emerges When Individuals Hold Divergent Values." *Academy of Management Journal* 57 (5): 1485–1512.

Bestill, Michele M., and Barry G. Rabe. 2009. "Climate Change and Multi-level Governance: The Evolving State and Local Roles." In *Toward Sustainable Communities*, edited by Daniel A. Mazmanian and Michael E. Kraft, 201–226. Cambridge, MA: MIT Press.

Birkland, Thomas A. 1997. *After Disaster: Agenda Setting, Public Policy, and Focusing Events*. Washington, DC: Georgetown University Press.

Birkland, Thomas A. 2006. *Lessons of Disaster: Policy Change After Catastrophic Events*. Washington, DC: Georgetown University Press.

Bruce, Christopher. 2006. "Modeling the Environmental Collaboration Process: A Deductive Approach." *Ecological Economics* 59 (3): 275–286.

Bryner, Gary, and Robert J. Duffy. 2012. *Integrating Climate, Energy, and Air Pollution Policies*. Cambridge: MIT Press.

Casado-Perez, Vanessa, Bruce E. Cain, Iris Hui, Coral Abbott, Kaley Doson, and Shane Lebow. 2015. "All Over the Map: The Diversity of Western Water Plans." *California Journal of Politics and Policy* 7 (2). http://dx.doi.org/10.5070/P2cjpp7225762.

Chan, Kwan Nok, and Shuang Zhao. 2016. "Punctuated Equilibrium and the Information Disadvantage of Authoritarianism: Evidence From the People's Republic of China." *Policy Studies Journal* 44 (2): 134–155.

Cox, Susan Jane Buck. 1985. "No Tragedy of the Commons." *Environmental Ethics* 7 (1): 49–61.

Creswell, John W., and Cheryl N. Poth. 2018. *Qualitative Inquiry & Research Design: Choosing Among Five Approaches*. 4th ed. Thousand Oaks: Sage Publications Inc.

Davis, Charles. 2012. "The Politics of 'Fracking': Regulating Natural Gas Drilling Practices in Colorado and Texas." *Review of Policy Research* 29 (2): 177–191. Accessed November 29, 2019. https://onlinelibrary.wiley.com/doi/abs/10.1111/j.1541-1338.2011.00547.x.

CVPP. 2022. "Central Valley Flood Protection Plan." https://water.ca.gov/-/media/DWR-Website/Web-Pages/Programs/Flood-Management/Flood-Planning-and-Studies/Central-Valley-Flood-Protection-Plan/Files/CVFPP-Updates/2022/Central_Valley_Flood_Protection_Plan_Update_2022_ADOPTED.pdf

Downs, Anthony. 1972. *Inside Bureaucracy*. Boston: Little, Brown.

EPA. 2022. *Environmental Protection Agency*. Accessed March 20, 2022. www.epa.gov.

Faulkner, Bill. 2001. "Towards a Framework for Tourism Disaster Management." *Tourism Management* 22 (2): 135–147.

Follett, Mary Parker. 2017. "The Giving of Orders." In *Classics of Public Administration*, edited by Jay M. Shafritz and Albert C. Hyde. Boston: Cengage Learning.

Geddes, Barbara. 2003. *Paradigms and Sand Castles*. Ann Arbor: The University of Michigan Press.

Gioia, Dennis A., and Henry P. Sims Jr. 1986. *The Thinking Organization*. London: Jossey-Bass Inc.

Greif, A., 1998. "Historical and Comparative Institutional Analysis." *The American Economic Review* 88 (2): 80–84.

Hanemann, Michael, and Caitlin Dyckman. 2009. "The San Francisco Bay-Delta: A Failure of Decision-Making Capacity." *Environmental Science & Policy* 12 (6): 710–725.

Hardin, Garrett. "Extensions of" the Tragedy of the Commons." *Science* 280 (5364): 682–683.

Healy, Andrew, and Neil Malhorta. 2009. "Myopic Voters and Natural Disaster Policy." *American Political Science Review* 103 (3): 387–406.

Heclo, Hugh. 1978. Issue Networks and the Executive Establishment." In *The New American Political System*, edited by A. King, 87–124. Washington, DC: American Enterprise Institute.

Herweg, Nicole, Nikolaos Zahariadis, and Reimut Zohlnhofer. 2018. "The Multiple Streams Framework: Foundations, Refinements, and Empirical Applications." In *Theories of the Policy Process*, edited by Christopher M. Weible and Paul A. Sabatier. Boulder, CO: Westview Press.

Homer-Dixon, Thomas. 2002. *The Ingenuity Gap: Facing the Economic, Environmental, and Other Challenges of an Increasingly Complex and Unpredictable Future*. London: Vintage.

Hopkins, Vincent. 2016. "Institutions, Incentives, and Policy Entrepreneurship." *Policy Studies Journal* 44 (3): 332–348.

Howlett, Michael. 2014. "Why are Policy Innovations Rare and So Often Negative? Blame Avoidance and Problem Denial in Climate Change Policy-making." *Global Environmental Change* 29: 395–403.

Jaques, E., 1990. In Praise of Hierarchies. *Harvard Business Review* 10 (1): 38–57.

Khalil, Elias L. 1995. "Organizations Versus Institutions." *Journal of Institutional and Theoretical Economics (jite)/Zeitschrift für die gesamte Staatswissenschaft* 151 (3): 445–466.

King, Gary, Robert Keohane Keohane, and Sidney Verba. 1994. *Designing Social Inquiry: Scientific Inference in Qualitative Research*. Princeton: Princeton University Press.

Kingdon, John. A. 1995. *Agendas, Alternatives and Public Policies*. London: Harper Collins. Accessed December 10, 2017.

Klyza, Christopher McGrory, and David J. Sousa. 2013. *American Environmental Policy: Beyond Gridlock*. Cambridge, MA: MIT Press.

Lam, Wai Fung, and Chung Yuan Chiu. 2016. "Institutional Nesting and Robustness of Self-governance: the Adaptation of Irrigation Systems in Taiwan." *International Journal of the Commons* 10 (2): 953–981. http://doi.org/10.18352/ijc.638.

Liden, Robert C., Sandy J. Wayne, Chenwei Liao, and Jeremy D. Meuser. 2014. "Servant Leadership and Serving Culture: Influence on Individual and Unit Performance." *Academy of Management Journal* 57 (5): 1434–52.

Lindblom, Charles E. "Still Muddling, Not Yet Through 1979. "*Public Administration Review* 39 (6): 517–526.

Lubell, Mark, Jack Mewhirter, and Ramiro Berardo. 2020. "The Origins of Conflict in Polycentric Governance Systems." *Public Administration Review* 80: 222–233. https://doi.org/10.1111/puar.13159.

March, James G., and Johan P. Olsen. 1989. *Rediscovering Institutions: The Organizational Basis of Politics*. New York: Free Press.

Mazmanian, Daniel A., and Michael E. Kraft. 2009. *Toward Sustainable Communities: Transition and Transformations in Environmental Policy.* Cambridge, MA: The MIT Press.

McAdams, Doug, Sidney Tarrow, and Charles Tilly. 2001. "Mobilization in Comparative Perspective." In *Dynamics of Contention,* 91–123. Cambridge: Cambridge University Press. https://doi.org/10.1017/CBO9780511805431.005.

McCray, Lawrence, and Kenneth A. Oye. 2007. *Adaptation and Anticipation: Learning From Policy Experience.* Boston: Center for International Studies, Massachusetts Institute of Technology.

McLaughlin, Milbrey W. 1987. "Learning From Experience: Lessons From Policy Implementation." *Educational Evaluation and Policy Analysis* 9 (2): 171–178. Accessed October 14, 2019. https://journals.sagepub.com/doi/abs/10.3102/01623737009002171.

Meijerink, Sander. 2005. "Understanding Policy Stability and Change: The Interplay of Advocacy Coalitions and Epistemic Communities, Windows of Opportunity, and Dutch Coastal Flooding Policy 1945–2003." *Journal of European Public Policy* 12 (6): 1060–1077.

Meinzen-Dick, Ruth. 2007. "Beyond Panaceas in Water Institutions." *Proceedings of the National Academy of Sciences* 104 (39): 15200–15205. https://doi.org/10.1073/pnas.0702296104.

Mintrom, Michael. 1997. "Policy Entrepreneurs and the Diffusion of Innovation." *American Journal of Political Science* 41 (3): 738–770.

Morris, John C. 2022. *Clean Water Policy and State Choice: Promise and Performance in the Water Quality Act.* New York: Cambridge University Press.

Morris, John C., William A. Gibson, William M. Leavitt, and Shana C. Jones. 2013. *The Case for Grassroots Collaboration: Social Capital and Ecosystem Restoration at the Local Level.* Lanham, MD: Lexington Books.

Mullin, Megan. 2009. *Governing the Tap: Special District Governances and the New Local Politics of Water.* Cambridge, MA: The MIT Press.

North, Douglass C. 1990. *Institutions, Institutional Change, and Economic Performance.* New York: Cambridge University Press.

Olson Jr., Mancur. 1965. *The Logic of Collection Action.* Boston: Harvard University Press.

Ostrom, Elinor. 1990. *Governing the Commons.* Cambridge: Cambridge University Press.

Ostrom, Elinor. 2002. "Common-pool Resources and Institutions: Toward a Revised Theory." In *Handbook of Agricultural Economics,* edited by Bruce L. Gardner and Gordon C. Rausser, Vol. 2, 1315–1339. Amsterdam: Elsevier.

Ostrom, Elinor. 2008. "Tragedy of the commons." In *The New Palgrave Dictionary of Economics,* edited by Steven N. Durlauf and Lawrence E. Blume, Vol. 2. Hoboken, NJ: Palgrave Macmillan.

Parks, Robbie, Megan McLaren, Ralf Toumi, and Ulrike Rivett. 2019. "Experiences and Lessons in Managing Water From Cape Town." *Grantham Institute* (Imperial College London) 29.

Petersen, Karen K. 2009. "Revisiting Downs' Issue-Attention Cycle: International Terrorism and US Public Opinion." *Journal of Strategic Security* 2 (4): 1–16.

Pierson, Paul, and Theda Skocpol. 2002. "Historical Institutionalism in Contemporary Political Science." *Political Science: The State of the Discipline* 3 (1): 1–32.

Press, Daniel, and Daniel A. Mazmanian. 2013. "Toward Sustainable Production: Finding Workable Strategies for Government and Industry." In *Environmental Policy: New Directions for the 21st Century*. 8th ed., edited by Norman J. Vig and Michael E. Kraft, 230–254. Los Angeles: CQ Press.

Quattrone, George A., and Amos Tversky. 1988. "Contrasting Rational and Psychological Analyses of Political Choice." *The American Political Science Review* 82 (3): 719–736.

Regmi, Ashok Raj. 2008. "Self-Governance in Farmer-Managed Irrigation Systems in Nepal." *Journal of Developments in Sustainable Agriculture* 3: 20–27.

Ross, Mark Howard. 2009. "Culture in Comparative Political Analysis." In *Comparative Politics: Rationality, Culture, and Structure*, edited by Mark Irving Lichbach and Alan S. Zuckerman, 134–161. Cambridge: Cambridge University Press.

Sabatier, Paul A. 1999. *Theories of the Policy Process*. Boulder, CO: Westview Press.

Scheberle, Denise. 2004. *Federalism and Environmental Policy*. 2nd ed. Washington, DC: Georgetown University Press.

Scott, W. Richard. 1995. *Institutions and Organizations*. Thousand Oaks: SAGE Publications Inc.

Seawright, Jason, and John Gerring. 2008. "Case Selection Techniques in Case Study Research: A Menu of Qualitative and Quantitative Options." *Political Research Quarterly*: 294–308.

Shanahan, Elizabeth A., Eric D. Raile, Kate A. French, and Jamie McEvoy. 2018. "Bounded Stories." *Policy Studies Journal* 46 (4): 922–948.

Shanahan, Elizabeth A., Michael D. Jones, Mark K. McBeth, and Ross R. Lane. 2013. "An Angel on the wind: How Heroic Policy Narratives Shape Policy Realities." *Policy Studies Journal* 41 (3): 453–483.

Simon, Herbert A. 1957. *Models of Man*, 196–297. New York: Wiley.

Spillane, James P., Brian J. Reiser, and Todd Reimer. 2002. "Policy Implementation and Cognition: Reframing and Refocusing Implementation Research." *Review of Educational Research* 72 (3): 387–431.

Steelman, Toddi A. 2010. *Implementing Innovation: Fostering Enduring Change in Environmental and Natural Resource Governance*. Washington, DC: Georgetown University Press.

Stone, Deborah A. 1997. *Policy Paradox: The Art of Political Decision Making*. New York: Norton.

Sutcliffe, Kathleen, M. and Timothy J. Vogus. 2003. "Organizing for Resilience." In *Positive Organizational Scholarship: Foundations of a New Discipline*, edited by Kim S. Cameron, Jane E. Duton, and Robert E. Quinn, 94–110. San Francisco: Berrett-Koehler.

Thelen, Kathleen. 1999. "Historical Institutionalism in Comparative Politics." *Annual Review of Political Science* 2 (1): 369–404.

Travis, Rick, Elizabeth D. Morris, and John C. Morris. 2004. "State Implementation of Federal Environmental Policy: Explaining Leveraging in the Clean Water State Revolving Fund." *Policy Studies Journal* 32 (3): 441–460.

Tyler, Stephen, and Marcus Moench. 2012. "A Framework for Urban Climate Resilience." *Climate and Development* 4 (4): 311–326.

Vig, Norman J., and Michael E. Kraft, eds. 2012. *Environmental Policy: New Directions for the Twenty-First Century.* 8th ed. New York: Sage.

Vogus, Timothy J., and Kathleen, M. Sutcliffe. 2007. "Organizational Resilience: Towards a Theory and Research Agenda." In *IEEE International Conference on Systems, Man, and Cybernetics*, 3418–3422. Montréal: IEEE.

Watts, J., Ronald Mackay, Douglas Horton, Andrew John Hall, Boru Douthwaite, Robert Chambers, and Anne S. Acosta. 2007. *Institutional Learning and Change: An Introduction.* Rome: Institutional Learning and Change Initiative.

Weatherley, Richard, and Michael Lipsky. 1977. "Street-Level Bureaucrats and Institutional Innovation: Implementing Special-Education Reform." *Harvard Educational Review* 47 (2): 171–197. Accessed October 14, 2019. https://hepgjournals.org/doi/abs/10.17763/haer.47.2.v870r1v16786270x.

Weaver, R. Kent. 1986. "The Politics of Blame Avoidance." *Journal of Public Policy* 6 (4): 371–398. https://doi.org/10.1017/S0143814X00004219.

Weible, Christopher M., Paul A. Sabatier, Hank C. Jenkins-Smith, Daniel Nohrstedt, Adam Douglas Henry, and Peter DeLeon. 2011. "A Quarter Century of the Advocacy Coalition Framework: An Introduction to the Special Issue." *Policy Studies Journal* 39 (3): 349–360.

Weible, Christopher M., and Tanya Heikkila. 2017. "Policy Conflict Framework." *Policy Sciences* 50 (1): 23–40.

Weick, Karl E. 1979. "Cognitive Processes in Organizations." *Research in Organizational Behavior* 1: 41–74.

Weick, Karl E. 1984. "Small Wins: Redefining the Scale of Social Problems." *American Psychologist* 39 (1): 40–49.

Welch, Susan, and Kay Thompson. 1980. "The Impact of Federal Incentives on State Policy Innovation." *American Journal of Political Science* 24 (4): 715–729.

Wildavsky, Aaron. 1988. *Searching for Safety,* Vol. 9. New Brunswick, NJ: Transaction Books. Accessed February 12, 2020.

Yin, Robert K. 2016. *Qualitative Research From Start to Finish.* 2nd ed. New York: The Guilford Press.

3 Texas

Scarcity and Cooperation

There are only a few topics that evoke an instinctual, emotional, and passionate response from Texans—football, traffic, and water all come to mind. Repeated droughts and floods, with an ever-present eye on future rain forecasts, has etched the topic of drought into the psyche of many Texans. Moreover, many in the state have a strong place-based attachment to the land (and to the state), which forms the key pieces of their identity (Miller 2020). Miller (2020) notes that Texas is a dry and arid land, but that makes the sense of surviving there all the more rewarding. This is the irony of water and drought in the state; Texans have built a mythology of conquering the dry land with a culture that appeals to outsiders, but as the population in Texas continues to boom alongside climate change, state and substate water planners are identifying new ways to provide services to an already stressed water environment (Miller 2020).

Background

Prior to accounting for policy and innovation within Texas's drought strategies, it is prudent to outline the terrain as applied to Texas (Table 3.1). Of note in Table 3.1 is that Texas faces a problematic confluence of stressors. In short, the state's fast-growing urban centers, and overall population growth, are increasingly demanding more water. Meanwhile, groundwater supplies are dwindling and exacerbating the state's water challenges in the future alongside the ever-present risk of short- and long-term droughts.

Texas's political ideology is best described as conservative. Yet, when it comes to clean energy and the water, the state is something akin to a leader. During the Obama administration, Texas was the bulwark of conservative opposition to federal environmental policies (Miller 2020, 195). Texas fought to delay several environmental policies, such as fracking restrictions and clean power policies. Simultaneously, Texas has invested more money into clean energy and water management than most states, currently producing almost one quarter of the entire country's wind energy (U.S. Energy Information Administration 2022).

DOI: 10.4324/9781003498537-3

Table 3.1 Scope and Scale of Water in Texas

Variable	Explanation
Source	Groundwater meets 60% of the state's needs, whereas surface water accounts for the remaining 40%
Aquifers (Groundwater)	"Texas is home to nine major aquifers that supply much of its ground water, including the Ogallala-High Plains Aquifer that stretches beneath eight states. Together, the Ogallala and Rita Blanca aquifers supply nearly 4.2 million acre-feet of water per year. The Gulf Coast Aquifer, stretching from Florida through Texas to Mexico, supplies 54 Texas counties with nearly 1.4 million acre-feet per year."
Surface Waters	Texas's surface water sources consist of 15 major rivers, 188 major reservoirs, seven major estuaries, eight coastal basins, and the Gulf of Mexico. Surface water abundance generally matches precipitation trends in Texas, though precipitation varies across the large expanse of the state, with the eastern region receiving far more than the western region.
Users/Population	• "Municipal needs such as residential water use account for about 27 percent of Texas's water demand, but that share is expected to grow dramatically over the coming decades as population increases. • Texas's population grew by more than 7 percent between 2010 and 2014 and is projected to increase by 82 percent between 2010 and 2060 to 46.3 million people."
Forecast	• Providing water for such substantial population growth will require a combination of additional water supplies and increased conservation. By 2060, Texas's water demand is projected to increase from 18 million acre-feet per year in 2010 to 22 million acre-feet per year. • Existing water supplies are projected to drop by 10 percent, from 17.0 million acre-feet in 2010 to 15.3 million acre-feet in 2060. This includes the projected depletion of the Ogallala Aquifer's supplies, which are expected to decline by about 2 million acre-feet per year, and the Gulf Coast Aquifer, which is expected to decline by about 210,000 acre-feet per year.
Vulnerability	• Based on daily per capita water availability, 10 urban areas in Texas are at medium or high vulnerability to water shortages, including San Antonio, El Paso, Dallas, and Austin.

Source: EPA 2016

Texas is often described as a mix of individual and traditional values (Elazar 1966). As originally written: In political culture, this corresponds to government control at the lowest level with limited impact and a push towards economic prioritization over other social priorities. This political philosophy is the same verbiage and prioritization seen in Texas environmental policies, illustrating a congruency between larger Texas culture and environmental governance. Water policy does have a wrinkle not found in other environmental areas though. The constant threat of drought has created a well-funded, hyper-focused policy community that attempts to maintain the balance between public need and personal rights. As one Texas water policy expert noted: "Texas is oddly progressive but based on local control and decentralization (Interviewee P 2022)." The question facing Texas is whether this balance will be enough to counteract future droughts.

Water Planning in Texas

In Texas, all water planning is based upon the "drought of record" or worst drought in Texas history, 1950–1957. At 77 months, this widespread drought left 253 (out of 254) counties as disaster areas and left a profound effect on the psyche of Texans in relation to water (TWDB 2012). The 2010–2014 drought brought the worst of Texans' fears back to the surface. While shorter in duration than the 1950s era drought of record, this span featured the single driest year in Texas history. For a 12-month period in 2011, Texas received an average of 10.86 inches of rain versus the historical baseline of 27.02 inches (NOAA 2015). Some towns, such as Spicewood Beach, were forced to truck in water daily, restricting households to only 50 gallons per day (de Melker 2012). In fact, some policy experts contend that the 2011 drought is the new drought of record as many water plans across the state were built more for duration rather than intensity. As a result, many communities and water systems caught off guard by what record-heat would do to their water planning assumptions (Interviewee P 2022).

During the 2010–2014 drought, the Texas Water Development Board (TWDB)[1] gathered the 16 water regions' plans to coordinate drought contingencies and mitigation strategies within a state framework. Once collected, regulators at the TWDB discovered that many of the regional water plans had no drought contingencies and that statewide coordination would require planning from scratch (TWDB 2017, 2019). Despite a statement in the 2012 State Water Plan that regions should "facilitate preparation for and response to drought conditions (235)," this drought brought to light how underdeveloped (if developed at all) many of the regional drought plans were.

Historically, lawmakers' primary policy response to water shortage in Texas was to build more reservoirs. From 1900 to 1959, the state supported the construction of approximately 100 lakes and reservoirs for water storage. Another 48 reservoirs were built in the 1960s, in response to the drought of record, but the rate of construction has slowed drastically since (Texas Almanac 2019). The 1984 State Water Plan proposed 44 reservoirs to meet future water supply needs, of which, 15 were constructed. More recently, in the 2002 State Water Plan, a total of 18 reservoirs were recommended with only one completed and one currently under construction. The slowdown in reservoir construction is due, in part, to several factors including difficulties in identifying viable sites for the construction of major reservoirs, permits are more difficult to obtain due to environmental concerns, and the cost of construction has exceeded the rate of inflation (Texas Living Water Project 2022). The recently opened Bois d'Arc Creek Reservoir, for example, came at a cost of about $1.6 billion, and it is the first major reservoir built in almost 30 years in Texas, taking more than 20 years from planning to water delivery into Dallas.

As noted in Table 3.1, the state finds itself in a precarious situation. Demand across the state is growing; the state requires an additional 1.5 million acre-feet (AF) of water. Moreover, groundwater resources are being depleted. Current studies predict that water management strategies, such as conservation alone, would not be enough to meet the future needs of Texans. The conventional solution, the construction of major reservoirs, is increasingly difficult (TWDB 2012), thus, setting the stage for innovation.

Environmental Variables

Water Distribution and Infrastructure

Texas's geography is dominated by flat plains and crossing rivers from high elevation to the west and to the southeastern coastline. Other than the Guadalupe mountains along the New Mexico border, Texas has little mountainous terrain. Western and Central Texas are considered arid, with increasing annual rainfall and sustained natural trees in the eastern stretches of the state. The lack of mountains means that Texas is almost entirely dependent upon rivers for water, most of which originate along the continental divide in New Mexico and Colorado, with small indigenous rivers from the central Texas Hill Country. Additionally, Texas does not have the luxury of internal snowpack seen in much of the West, again forcing the state to rely almost exclusively on a few rivers flowing throughout the state. There is only one natural lake in Texas; however, over 200 manmade lakes and reservoirs are used to control flooding and provide water resources for the state.

From approximately 2019 to 2022, Texas water usage consisted of an average of 54 percent groundwater, 43 percent surface water, and 2 percent reused water (TWDB 2019). This is a significant improvement from the height of the 2011 drought, where 62 percent of all Texas water came from groundwater, as compared to only 38 percent from surface water. Texas water management efforts have also benefited from the incorporation of modern water techniques, such as reused water, which has received increased attention in the 2017 State Water Plan (TWDB 2017, 2019). As such, Texas uses an average of 14 million AF a year, with 6.6 of that coming from surface water (TWDB 2019). Of this total amount,

- 34 percent is for urban usage;
- 14 percent is for commercial usage;
- 52 percent is for agriculture and livestock.

These percentages are largely stable. There was an uptick in irrigation at the height of the drought, but this is mostly a result of irrigation from groundwater, which was overpumped to compensate for the heat and lack of rainfall (TWDB 2019; Hill and Whitehead 2019).

Environmentally, the distribution of water is heavily weighted toward the east. Figure 3.1 shows the annual rainfall distribution throughout the state, highlighting the arid desert in the far west compared to the humid eastern border. The black line on the map shows the I-35 corridor, connecting from north to south: Dallas/Ft Worth, Waco, Austin, and eventually San Antonio. In fact, five of Texas's seven largest cities are located within this corridor, excluding Houston and El Paso. The location of these major cities in Texas is the cause of tension within the state as many of these cities have only one direction to look future water development, toward the less-populated, but more water-rich eastern woodlands, as shown in Figure 3.1 (Interviewee O 2022). In fact, West Texas receives approximately 8 inches of annual rainfall, whereas areas in southeastern Texas receive around 56 inches annually.

As previously stated, Texas does not have the large river systems seen in the other states present in this book. By comparison, the Central Valley Project in California distributes an average of 9.5 million AF a year, more than the top five rivers in Texas combined. Alabama has an estimated 102 million

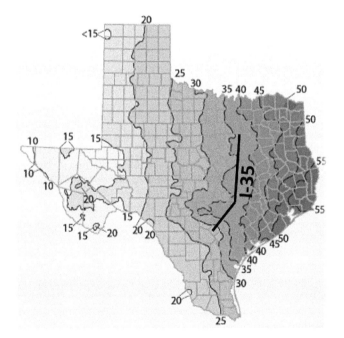

Figure 3.1 Annual Rainfall in Texas.

Source: Texas Water Development Board.

Figure 3.2 Aquifers in Texas.

Source: TWDB 2022

AF of water running through the state annually, with the lowest drought-level flow of the Alabama River at 6.5 million AF, roughly equal to the top three Texas rivers on an average year. As a result, Texas relies heavily on aquifers for urban supply. TWDB estimates that the state has about 14.2 million AF of usable aquifer storage in the state, although that is biased toward a few large aquifers. The Ogallala in the panhandle, the Edwards around San Antonio, and the Gulf Coast in the southeast have over one-half of the aquifer storage in the state (TWDB 2022), as shown in Figure 3.2.

There is no large-scale infrastructure to distribute water throughout the state, which pushes planning from the state level, as seen in California, down to the regional and local levels. This has driven the Texas water battles, from a local to a regional-based planning system, where coordination happens on local level and is filtered up to the state for long-term prioritization.

Structural Variables

Rules and Communication

Texas water law is the foundation of a multitude of rules and governing structures that shape state water planning efforts as well as the development of organizations designed to enforce such laws. Texas has two major water-related
bodies, the Texas Commission on Environmental Quality (TCEQ) and the
TWDB. TCEQ is primarily a regulatory agency, controlling permits and li-
censes, whereas TWDB is primarily a research agency that gathers information, coordinates substate planning, and makes policy recommendations to
state lawmakers (Texas Commission on Environmental Quality 2023).

Water planning in Texas utilizes a "bottom-up" approach, which begins
with local-level planning and coordination. Local water planners are empow-
ered to monitor their geographical areas and create long-term forecasts that
are distributed to the 16 water regions who are tasked with (1) planning for
drought of record conditions; (2) conducting evaluations of future water demands, existing supplies, potential shortages, and feasible water management
strategies; and (3) reporting the associated data (by decade and broken down
geographically) over a 50-year planning period (TWDB 2017). The regional
water plans aggregate much of this data and then provide a city-by-city de-
scription (only those jurisdictions within their regions) of current and future
threats to water, individual community water sources, and triggers for drought
response (TWDB 2019).

Twice each decade the TWDB compiles the 16 regional plans and ranks
priorities based upon state-wide needs and requirements to produce the State
Water Plan (SWP). The SWP coordinates district-level plans as well as prom-
ulgates a series of recommendations to the Texas Legislature. The latest iteration of this plan has been published online to provide interactive access to
all pertinent data, to include population estimates, water use categories, and
projected water demands (TWDB 2017).

Like most states, Texas Water Code (TWC n.d.) divides water into two categories: surface water and groundwater. Surface water in Texas is owned by
the state, held in a trust for the public (TWC 11.021, 11.01235), and is managed by TCEQ. In order to use state-owned water, a person must file a permit
and obtain a "water right" from the state, utilizing the "first in time, first in
right" allocation structure. In times of shortage, a senior water right holder
receives all the water to which he or she is entitled before a junior user will
get any water, unless an emergency situation develops, at which time water
can be arbitrated between the parties (Dowell Lashmet 2018). There are minor
exceptions to a permit for riparian owners, such as domestic and livestock
use, wildlife management, and emergency situations; otherwise, a permit is
required for water usage (Texas Water Code [TWC] 11.142).

Groundwater in Texas is more complicated. Groundwater is the property of the landowner, who has the authority to use all the water they can capture, regardless of the effect on neighbors (TAMU 2014). Unlike most other states, Texas courts have denied "reasonable use" as long as water is not maliciously taken for the purpose of causing injury to a neighbor or willfully wasted (Dowell-Lashmet 2018; TAMU 2014). That said, the state prefers regional management of groundwater through Groundwater Conservation Districts (GCDs), which now control most of the groundwater within the state and have developed specific regulations for landowners within their boundaries.

GCDs have been a part of Texas water planning for over 50 years, but since the 1980s, the state has strongly encouraged their use (TAMU 2022). To form a GCD, voters in the region must approve its creation, establish leaders, and fund the body through taxes. Once this is completed, the legislature approves the district. Due to the political backing required for their creation, these districts are usually located along county lines, whereas groundwater extends beyond political boundaries, often requiring multiple districts to agree on a common use plan of groundwater (Interviewee O 2022).

Texas districts may set sustainability or consumption goals, such mandates are not incorporated into state water policy code like is seen in other states' groundwater laws (Interviewee P 2022). Case law findings and private property precedents maintain water as an owner's right, so any change to these policies will require legislative action. Texas is fortunate to have diverse groundwater resources, but the premise of private property ownership has made it difficult to establish caps for regulatory agencies (Interviewee P 2022). In short, groundwater conservation districts are certainly a step in the direction of coordination and sustainability, but the lack of coordination between the groups, each with their own policies and procedures, is likely to be a point of conflict and innovation impediment in the future (Interviewee O 2022).

Communication in the state appears to be open and direct, with an emphasis on future drought preparedness. As will be discussed later, the major urban areas have regular public meetings with channels for communication throughout the process. The Texas water planning regions hold regular open meetings with direct inputs from impacted communities. In interviews with participants throughout the policy chain, respondents were each clear about the water planning process and their role in it; from local planners to regional board members, the rules of Texas water are straightforward and well-understood, at least among experts (Interviewee P 2022; Interviewee V 2022; Interviewee O 2022).

An interesting wrinkle for planning is that the individual communities often make a distinction between policies and projects. The City of Dallas, for example, does not recommend large-scale water policies up to the state; however, it does recommend individual projects, such as reservoirs, for delivering water to its customers (Interviewee O 2022). In this sense, planners feel as though they are simply following the state's water guidance and providing inputs on how to achieve state-supported goals (Interviewee O 2022). This places much of conceptual and long-term planning at the regional level based on detailed inputs from individual communities. The state then takes the regional policy recommendations and stacks them according to priority and necessity, which feeds into the State Water Plan, which is then shared with the legislature. According to interviewees, there is open communication from the regional level through to the state legislature, with direct inputs from communities.

One of the reasons for the clear communication in the state is the simple fact that the state is now on its 11th iteration of the state water plan. There were water plans in Texas before the 1950s drought, but nothing long-term. The 1961 State Water Plan was the first modern planning attempt and resulted in Texas's current water infrastructure (Interviewee O 2022). Updates were somewhat sporadic for years until the 1997 Texas Senate Bill 1 (SB-1) created the modern water planning process. Since the early 1990s, the state has averaged a new water plan every five years, meaning that the rules and communications between the levels of planning have become standardized and familiar to stakeholders. While rules can often lead to rigidity in planning, the constant threat of drought in the state keeps the lanes of communication open throughout the planning chain.

A vexing communication challenge arose during the 2011 drought. As conditions worsened, many communities discovered that they had promulgated different measures of drought-related definitions, action plans/responses, and solutions (Interviewee C 2022). Cities, for example, enacted different definitions for what constitutes the varying drought stages, especially what would equate to Stage 1 conditions, as well as different requirements and restrictions for their citizens. It became particularly confusing when news stations reported restrictions for Dallas that conflicted with Ft. Worth requirements (Interviewee C 2022). The cities soon realized that in the interest of legitimacy and clarity for their populations, the clear communication of consistent restrictions within major areas was necessary. As a result, the DFW metroplex published new guidance in 2014 that sought to harmonize (or at least coordinate) responses across individual communities.

Incentives

The incentives for water innovation in Texas look drastically different than in other states. While some states set aside grant funding for municipalities, Texas law largely restricts the state's grant authority, as many communities either do not qualify for grants or do not have the individual expertise to apply for them (Interviewee P 2022). That said, Texas has heavily funded water research via direct funding. The standardized planning system means that communities can request water innovations directly from the state, as opposed to a grant-based system. Innovative policies, from water reuse to groundwater management, are equally encouraged from the state legislature. In fact, since 2016, Texas has spent an estimated $9 billion on water infrastructure (TCEQ 2022).

As for citizens, water experts across the state have mentioned the willingness of citizens to do their part in drought control (Interviewee O 2022; Interviewee P 2022; Interviewee V 2022). There will always be outliers, but for the most part, citizen response to agency requests has been positive. One water expert mentioned how when San Antonio asked citizens to cut back on water between 2011 and 2014, the per capita usage dropped more than planners had expected (Interviewee P 2022). More recent droughts, however, have enjoyed less citizen "buy-in." The city attributes the different levels of citizen support to patterns related to new home constructions. According to the city, homes built since 2011 tend to have more auto-irrigation, which uses up to 70 percent more water than "hose-drawn" homes. Citizens tend to turn the sprinkler systems on and forget about them, leading to a lower response than their previous planning assumptions (Interviewee P 2022). In this sense, the planning factors from just a few years ago had to be revised, as the city had to engage in directed communication at specific neighborhoods to engender the desired response.

Direct federal funding did/does not play a large role in the day-to-day water innovation at the state level in Texas. That said, there are projects within the state that have benefited from federal funding. As an example, the Tarrant regional water supply is seeking federal funding for flood control through the construction of natural flood barriers (Interviewee V 2022). Furthermore, some of the state environmental funding, such as the State Revolving Funds, was initially seeded from federal Environmental Protection Agency. Each of the interviewees responded that while federal funding was a part of Texas environmental policy, it was not a driver in the state's decision-making. For cities, federal funding can enable smaller projects (that the city would not normally prioritize), such as connecting rural houses to municipal water systems, that would not be cost-effective for the city (Interviewee O 2022).

Opening

Opening refers to the ability for citizen and interest groups to make their opinion known in state policy. Texas's water planning bodies are open to public comment and interaction, from the local level through the TWDB, which has been lauded by environmental groups (TWDB 2022):

> Sierra Club Lone Star Chapter and Hill Country Alliance commented that they have seen a marked improvement in the sophistication, credibility, and value of State Water Plans since the passage of Senate Bill 1 in 1997.

They, along with the National Wildlife Federation and Galveston Bay Foundation, applaud the development of the interactive State Water Plan and other data transparency efforts to view historical and projected water data.

As part of the last two water plans, the TWDB published 26 (2017) and 50 (2022) pages of public comment on the plan, organized by topics such as general comments, policy recommendations, drought response, climate change, and groundwater (TWDB 2022). The 2017 comment release even had a section on transparency, in which multiple commenters, including Wilson County Water, southeast of San Antonio, complained about the transparency within Region L, which was noted by the TWDB (2017, 5). Included in these pages of public comments are asterisks (*) that indicate when a comment resulted in a water plan change. It is important to highlight the diversity of comments in these documents including by interest groups such as the Sierra Club Lone Star Chapter, to by individual commenters and local advocacy groups. In the 2017 publication, TWDB responded directly to Sierra Club-specific comments, whereas the 2022 publication included 28 TWDB comments and responses to specific Sierra Club concerns.

On the local level, the commitment to public engagement is more varied. Some municipalities, for example, are intentional about soliciting input before submitting their plans while others are not. As an example, the city of Dallas shared its water plan forecasts to multiple environmental groups and collected their responses prior to publishing the plan (Interviewee O 2022). Even though the planners knew there would be disagreements, according to Interviewee O, sharing the documents helped establish trust between the groups. Moreover, it afforded the City the opportunity to explain the "why" behind the plans, even if the included groups disagreed on the "how" (Interviewee O 2022). Similarly, Ft. Worth formed a stakeholder group consisting of commercial, industrial, residential, and environmental actors to offer feedback to the city. After the first draft of the recent revision was completed, the city sent a copy of its plans to many of these stakeholders for feedback, before the plan was presented to the city council (Interviewee V 2022).

One policy expert described how the individual participants change depending on the planning level (Interviewee P 2022). On the local level, there are always environmental groups active but also private property advocates. At the regional level, larger manufacturing, trade, and environmental interest groups are more active, with less individual interaction. On the state level, often NGOs, such as the Texas Water Conservation Association, are the ones writing the actual policy to be implemented (Interviewee P 2022). In this sense, the incorporation of local planning into regional and state plans allows for those local-area activists to have a voice in state planning, even if they would be unable to participate directly.

Resistance

Resistance and conflict in environmental policy are a hallmark of the field, and in a state as large and environmental diverse as Texas, conflicts are inevitable. From 2011 to 2016, just over a thousand "environmental" lawsuits were filed in the state of Texas (Justia 2022). Of these lawsuits, 29 percent directly involved either the Texas Water Development Board or the Commission for Environmental Quality. The total number of environmental lawsuits is about one-quarter of California's over the same time span, but with a much higher participation rate for the state water entities.

One of the primary reasons for litigation in Texas involves land that has been set aside for future reservoir construction. To set aside land for a reservoir, privately owned land must be purchased from landowners or confiscated as "eminent domain." This also keeps these regions from being commercially developed, limiting growth in these areas. The second challenge for reservoirs is that these sites are generally not collocated with the water end-user (Interviewee O 2022). Within Texas, most major metro areas look to the east for water. As a result, the cities around the identified future reservoirs bear the brunt of the costs via restricted economic development opportunities and little to no benefit from the water when completed. Third, the permitting process for a single project can take 15–20 years, where "during that time when [federal] personnel turn over, we have to restart the process (Interviewee O 2022)." As a recent example, the Bois d'Arc Creek Reservoir northeast of Dallas is the first reservoir built in the state since the 1980s. The project took just over three years to build but 15 years to permit (Interviewee O 2022).

Other than reservoir construction, the state is relatively cooperative on water projects. One interesting difference between Texas and California is that the major water users are decoupled from supplies. As one planner stated: "I don't know of any parties that go after the same supply (Interviewee O 2022)." This is a marked difference from a state like California, where every drop of water is up for grabs from Sacramento to San Diego. Texas has invested billions of dollars to create an open water structure that has led to much more stability throughout the state (Interviewee P 2022).

Cultural Variables

Shocks

Major shocks often serve as the spark that drives innovation in policy. In Texas, shocks have shaped the psyche of the average citizen as much as the network of water policy actors. As previously stated, Texas has endured a series of shocks from the long-term drought to shorter, more-intense droughts. Officially, the "drought of record" was a seven-year drought in the 1950s;

however, recent droughts, serving as a shock, have elucidated the flaws and limitations in these planning assumptions.

The 2011 drought was the single driest year on record and captured the public attention. It garnered national headlines and was the first time the state enacted Stage 1 drought restrictions. As stated in Chapter 1 for many in Texas, drought is a constant fear and is burned into the psyche of Texans (Interviewee P 2022; Downs 1972). Yet, the shock from 2011 was less on the populace, who (at the time) understood their role in drought mitigation, but more for the water providers and planners. As an example, San Antonio's Edwards Aquifer, the primary source of water for the metropolitan area, lost about 40 percent of its volume in the hot and dry weather. At the start of 2011, the water level in the Hondo Well of the aquifer was at 711 feet, near its historical average of 718 feet. By June 21, it was 664 feet, a rapid drop in the aquifer availability. The water level in the aquifer did not get above 700 feet until May 15, 2016, five years after the drought began (Edwards Aquifer Authority 2022).

A topic mentioned repeatedly by policy experts was the fact that they can never know the true weakness of their planning until the plan is implemented (Interviewee V 2022; Interviewee P 2022). This has repeatedly been true for Texas. Just when planners feel as though they understand the problem, Mother Nature throws a curveball, and those old plans are no longer valid. That is the current state of Texas water planning, as a new and more intense drought has recently started (Interviewee V 2022; Interviewee P 2022). Although outside the scope of this research, the 2022 summer was a "flash" drought, with the hottest and driest summer on record (Interviewee C 2022). Currently, the same Edwards aquifer is at a six-month average of 651 feet, 63 feet below the historical average, and less than even the 2011 drought. While planners understood the dangers of prolonged droughts, reservoirs and aquifers appear to be more vulnerable to flash droughts than originally anticipated. Edwards was around 700 feet at the end of 2021 but went as low as 645 feet on August 6, 2022. This has planners struggling to understand drought mitigation moving forward and adjusting their drought triggers and assumptions (Interviewee P 2022).

Framing

One of the most difficult issues with environmental policy is defining the problem and framing it to further your agenda. In Texas, much of the debate centers on who is/what is to blame and the type of solutions that make the most sense. Environmental groups frame much of the water shortage as the result of poor infrastructure and flood mitigation. According to the National Wildlife Federation, Texas loses 572k acre feet per year, which would be enough to supply Ft. Worth, Austin, Lubbock, El Paso, and Laredo (Walker et al. 2022).

While this article is not peer-reviewed, it does highlight a very real problem of aging infrastructure in metropolitan regions such as Dallas and Houston. The primary argument is that if the state can fix the existing infrastructure, there is no need for other expensive solutions, such as reservoir construction. The state has countered this argument by saying that it has spent $9 billion on infrastructure upgrades in the past decade, so the problem is being addressed, but maybe not at the speed desired (TWDB 2012, 2017, 2019, 2022).

By comparison, the State Water Plan for 2022 frames the problem as a growth and supply issue. The strength of Texas's water planning is the city-by-city detail of supply and expected demands. Yet, even the best projections can be wildly off. The 1997 State Water Plan, for example, estimated the 2030 population for Region C (Dallas/Ft Worth) of 7.48 million people. Region C exceeded that estimate 12 years earlier than predicted, with current estimates of 8.86 million people by 2030 (TWDB 2022). The thought of restricting water taps in the future was one that was not amenable to the interviewees; unequivocally they felt sufficient water was available for a growing population, underscoring the push for reservoir construction as a buffer against limited water availability (Interviewee V 2022; Interviewee C 2022).

Throughout the period of study, consciousness of droughts has remained high for the average Texan. The 2011 drought again saw a spike in drought interest in Texas as the problem became national news. Interestingly though, interest in the story waned by early 2012, despite continued newspaper coverage, as shown in Figure 3.3. As the drought spiked in late 2011, news stories in the *Dallas Morning News* spiked and only slowly waned over the next two years to the pre-drought levels by mid-2013. Conversely, public interest in the

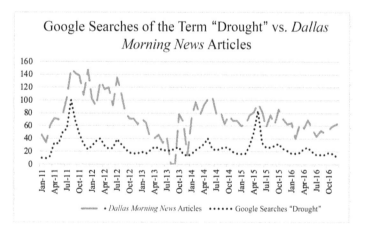

Figure 3.3 Google Searches of the Term "Drought" Versus *Dallas Morning News* Articles.

drought spiked and then quickly regressed to pre-drought numbers. The subsequent spike in Google searches in 2015 correlated with the catastrophic drought in California, again showing the relationship between local public attention and national events. There was a slight increase in *Dallas Morning News* articles covering California, but nothing compared to the increase of Google Trends.

Legitimacy

How states and interest groups legitimize their the proposed solution is indicative of their view of the problem. In Texas, there are two primary solutions, more storage or more efficiency. Environmental groups, from the Sierra Club to the National Wildlife Federation, do not view increased reservoir storage as a viable long-term solution, citing the environmental damage done by reservoirs (Texas Living Water Project 2022). Yet, there is general agreement between government, some in industry, and environmental groups on the need for greater water efficiency within the state. The Texas Living Water Project (2022), for example, lauds how the City of San Antonio, despite population growth, has maintained steady water withdrawals. Similarly, every major metropolitan area in Texas has maintained drought-stage restrictions on lawn watering, among other cutbacks (Interviewee C 2022, Interviewee P 2022). The environmental groups have been honest in their praise of these efforts, with the hope that these will be sufficient for the future water needs.

In other instances, legitimacy is achieved through the disclosure of information. The City of Ft. Worth is illustrative. Here, the city legitimizes water conservation efforts via multi-tiered billing. There are four tiers in Ft. Worth, and the more water used, the more expensive it becomes. The lowest tier for the city is below the cost of service, meaning that if each citizen kept their consumption to that rate, the city would pay more to deliver water than they received in fees (Interviewee V 2022). Here the city is incentivizing water savings by pushing water conservation to the end users, while punishing those who do not abide. Similarly, San Antonio has found that sending personalized mails to individual citizens is particularly effective. Instead of broad statements about citywide water usage, letters addressed to individuals about their water waste compared to neighbors has been effective for the high-volume users (Interviewee P 2022). While conservation is important, the state argues that conservation alone will not solve the water crisis and has sought to legitimize the expansion of water storage in the near future.

The 2022 State Water Plan has 20 potential reservoir locations identified that are the backbone of future growth, although permitting and litigation has slowed their development. According to TWDB, conservation is necessary for future growth, but not sufficient. The impact of recent flash droughts and more intense summers will likely not change the opinion of the state in their planning; however, interest groups argue that the recent high-intensity droughts were particularly detrimental to surface storage, causing rapid declines in reservoir levels (Douglas 2022). While these two sides each look to legitimize their own potential solutions to drought in the state, there is much more agreement on the importance of conservation than conflict over reservoirs.

Policy Innovations (2011–2016)

One important innovation learned from the 2011 drought was the necessity for common language and responsibilities within the regional planning process. Since 2011, the TWDB has taken a more proactive role in ensuring that regions have thoroughly planned for drought contingencies (TWDB 2017). While water is still managed on a regional level, those regions are held accountable at the state level, creating a more uniform and complimentary state water plan while still permitting local innovation.

As water scarcity increases, alternate sources of water will become more important. Water reuse is now specifically tracked and reported by municipalities, which is now far more encouraged than before, and desalinization is expected to play a more important role in Texas water management (Interviewee P 2022; Interviewee V 2022). Currently, the city of Corpus Christi is the only municipality actively researching desalinization; however, research into the conversion of brackish groundwater is expected to provide another source of water throughout the state in the future (TWDB 2019).

For future planning, the TWDB (2019) report on water conservation provides an outlook for water demand out to 2070. The report breaks down the future water management strategy, with 30 percent of additional future water coming from unused surface water, 27 percent coming from increased conservation, 13 percent from new reservoir construction, and the remainder from a collection of minor improvements, such as reuse and desalinization.

As Dallas, Ft. Worth, and San Antonio each learned from the 2011 drought, water planning is only as useful as the underlying assumptions. The 2011 drought, as well as the current 2022 drought, showed that water level assumptions were largely incorrect based on the older drought of record. Many of these changes have been incorporated, but this drought taught these municipalities the importance of even basic assumptions in water planning, such as population growth and water availability (Interviewee P 2022; Interviewee O 2022; Interviewee C 2022). As an example, the population boom in the 2010s due to the growing Austin tech sector, now rivaling Silicon Valley, was not accurately modeled in the water plans.

Similarly, assumptions of citizen performance were flawed, but in the other direction. Citizens were willing to cut back on their water usage, except for high-volume users that required direct intervention (Interviewee P 2022). One potential innovation here is the use of Advance Meter Infrastructure (AMI) "smart" water meters, which report a myriad of water usage details directly to the city. Many smaller cities in Texas incorporated this technology because of the 2011 drought, but some larger utilities are struggling with funding. San

Antonio has not been able to fully fund the project due to the high cost and the number of city water meters but is currently making the case for its implementation (Interviewee P 2022).

This points to perhaps the most effective innovation for Texas since the 2011 drought. The opportunity to enact their local and regional water drought plans, although it was due to the extreme nature of the drought, allowed the cities and regions to learn more about drought response and leverage "smart" technology. The state has a recurring planning process and has invested heavily in positive adaptations to crises (Moench and Tyler 2012). A large part of this is the investment in information. From digital meters to reservoir levels, the details of information for decision-making have risen significantly. With each recurring drought, organizations have developed new resources to capture important lessons, and the drought response plans have become more effective.

Conclusion

The state of Texas takes water very seriously. After each drought, the state takes stock of the lessons learned and adjusts current and future assumptions to better provide service to their citizens. From major metropolitan areas to small towns, the structure of the state water system seemingly allows for process improvement and innovation. There were no sweeping, state-wide innovations from the 2011 drought, but there were numerous small updates and improvements and significant innovations at the local level. This dynamic is largely a testament to the investment in water planning up to this point. The 2017 SWP was a step forward in regional planning detail and the increased water reuse and conservation, but otherwise it was a data-driven and improved iteration on the previous system.

Support for the Framework

This conclusion evaluates each proposition based upon findings from this case.

(1) States with highly detailed, frequently reviewed water plans (<10 years) will be less likely to enact non-incremental policy innovations in response to a new drought event.

Texas has established water policy culture and institutions. This water policy culture is well funded by the state and is indicative of a commitment to water planning. The 2011 drought did not result in any sweeping major changes for Texas but has seen significant substate innovations. It did highlight deficiencies in the policy process, which has been addressed in subsequent water plans. This case supports this statement with a caveat that substate innovation was a novel finding.

(2) States with distributive water infrastructure will experience a more competitive and fractured water political culture.

Texas does not have a distributive water system. Major cities are largely confined to their regional area for water sources. These metropolitan areas are not in competition for the same water sources, as such are largely cooperative throughout the state. There is some competition between these areas and smaller communities to the east for their water resources, but this is relatively minor. This case appears to support this statement. There appear, however, to be significant challenges associated with the costs and benefits of reservoir construction that has contributed to conflict.

(3) States with competitive environmental cultures will have less innovative policies.

Texas does not have a competitive environmental culture but does have top-down water policies. This case does not directly test this statement but will allow for comparison with other cases.

(4) States whose citizens have experienced major water shocks will have a more cohesive, innovative policy culture.

Texans have experienced successive major droughts in their history. The 2011 drought showed that citizens were largely cohesive in their drought response, exceeding water planners' assumptions. Similarly, per capita water use in major cities has steadily decreased over the past few decades. This case supports this statement.

Note

1 In 1957, state lawmakers established the Texas Water Development Board (TWDB) to maintain the viability of the state's water supply (among other resources). To do so, it manages, collects, and disseminates water-related data; coordinates and assists with regional water supply issues; and prepares the state's main water plan in conjunction with regional authorities (TWDB 2017).

References

de Melker, Saskia. 2012. *Two Texas Towns Run Out of Water*. March 20. www.pbs.org/newshour/science/science-jan-june12-texaswater_03-20.
Douglas, Erin. 2022. *Texas' Plan to Provide Water for a Growing Population Virtually Ignores Climate Change*. October 31. Accessed December 15, 2022. www.texastribune.org/2022/10/31/texas-water-plan-reservoirs-climate-change/.

Dowell-Lashmet, Tiffany. 2018. "Basics of Texas Water Law." *Texas A&M AgriLife Extension.* January. https://agrilife.org/texasaglaw/files/2018/01/Basics-of-Texas-Water-Law.pdf.

Downs, Anthony. 1972. "Up and Down with Ecology: The "Issue–Attention Cycle." *The Public Interest* 28: 38–51.

Edwards Aquifer Authority. 2022. *Historical Data.* Accessed December 16, 2022. www.edwardsaquifer.org/science-maps/aquifer-data/historical-data/.

Elazar, Daniel J. 1966. *American Federalism: A View from the States,* Vol. 61. New York: Crowell. Accessed November 30, 2019. https://cambridge.org/core/journals/american-political-science-review/article/american-federalism-a-view-from-the-states-by-elazar-daniel-j-new-york-thomas-y-crowell-1966-pp-xii-228–250/a45d4db4cd3ec0b738a59e8040b87b8f.

Environmental Protection Agency (EPA). 2016. "Saving Water in Texas." Accessed November 14, 2022. https://www.epa.gov/watersense.

Hill, Jason T., and Victoria Rose Whitehead. 2019. "Decoding Water Law: Ten Areas of Texas Water Law Every Ag Lawyer Should Know." *Texas A&M Journal of Property Law* 5: 449.

Interviewee C in discussion with the authors. July 29,2022.

Interviewee O in discussion with the authors. July 25, 2022.

Interviewee P in discussion with the authors. October 19, 2022.

Interviewee V in discussion with the authors. July 27, 2022.

Justia. 2022. *Justia Dockets and Filings.* Accessed September 5, 2022. https://dockets.justia.com/.

Miller, Kenneth. 2020. *Texas vs. California: A History of their Struggle for the Future of America.* New York: Oxford University Press.

Moench, Marcus, and Stephen Tyler. 2012. "A Framework for Urban Climate Resilience." *Climate and Development* 4: 4. Accessed February 12, 2020..

NOAA. 2015. *Global Climate Report—Annual 2015.* Washington, DC: National Oceanic and Atmospheric Administration.

Texas Almanac. 2019. *Lakes and Reservoirs.* https://texasalmanac.com/topics/environment/lakes-and-reservoirs.

Texas A&M University [TAMU]. 2014. *Texas Water Law.* https://texaswater.tamu.edu/water-law.

Texas A&M University [TAMU]. 2022. *Groundwater Conservation Districts.* Accessed December 19, 2022. https://texaswater.tamu.edu/groundwater/groundwater-conservation-districts.html.

Texas Commission on Environmental Quality [TCEQ]. 2022. "Water and Wastewater Funding Sources." www.tceq.texas.gov/assistance/water/water-and-wastewater-funding-sources

Texas Commission on Environmental Quality [TCEQ]. 2023. "Water." www.tceq.texas.gov/agency/water_main.html.

Texas Living Water Project. 2022. *Reservoirs Are Not the Future—Instead, Look to Water Conservation.* Accessed December 16, 2022. https://texaslivingwaters.org/state-and-regional-water-plan/reservoirs-are-not-the-future-instead-look-to-water-conservation/.

Texas Water Code [TWC]. n.d. "Texas Water Code." *TWC.* https://codes.findlaw.com/tx/water-code/water-sect-11-142/#:~:text=1427%2C%20§%201%3E-,(b)%20Without%20obtaining%20a%20permit%2C%20a%20person%20may%20construct,but%20not%20including%20fish%20farming.

TWDB. 2012. *2012 State Water Plan.* Austin: Texas Water Development Board.

TWDB. 2017. *2017 State Water Plan.* Austin: Texas Water Development Board.

TWDB. 2019. *Special Reports to the Texas Legislature.* Austin: Texas Water Development Board.

TWDB. 2022. *2022 State Water Plan.* Austin: Texas Water Development Board.

U.S. Energy Information Administration. 2022. *Electricity Data Browser.* November 6. Accessed November 6, 2022. www.eia.gov/electricity/data/browser/.

Walker, Jennifer, Alan Wyatt, Jonathan Seefeldt, Danielle Goshen, Meghan Bock, lan Johnston, Maya Black. 2022. *Hidden Reservoirs: Addressing Water Loss in Texas.* Austin, TX. Accessed December 16, 2022. https://texaslivingwaters.org/wp-content/uploads/Hidden-Reservoirs-Addressing-Water-Loss-in-Texas.pdf.

4 California

A Battle Over Historic Water Rights

The state of California is one of the most environmentally conscious regions on the planet, rooted in its rich geography and natural wonders. The Sierra Club, perhaps the most well-known of all environmental groups, was founded in San Francisco in 1892 by the naturalist John Muir (Miller 2020; Cohen 1988). In order to protect the beauty and resources of the Sierra Nevada, and specifically the Yosemite area, from expanding populations, Muir organized and activated like-minded citizens who lobbied to protect the California environment at a time of booming expansion and consumption of resources. While Muir won battles like Yosemite and lost battles like Hetch Hetchy, the struggle between the populations on the coast and the natural wonder in the mountains set the stage for environmental policy in California.

Although California is one of the most politically progressive states in the United States, the reality of providing services to almost 40 million residents puts the state in a unique position. While the aforementioned Hetch Hetchy reservoir flooded one of the most pristine and beautiful valleys on earth, it did so to provide water stability for the growing city of San Francisco (Miller 2020). California has water, just not where the population lives.

California is also plagued by another problem rarely found in drought-prone states. The state must deal with both sides of the water spectrum: extreme droughts and extreme floods. On average, 50 percent of California's annual precipitation falls between five and 15 days each year (Dettinger 2011). This provides a rare challenge for the state, as it has water storage capacity, but the storage must be able to absorb a large amount of water if an atmospheric river event happens during the winter (Interviewee T 2022). In this sense, California water planners must often make compromises on storage solutions, stuck between the danger of not storing enough water for the dry summer months but leaving enough room to absorb the wet winter storms.

One of the problems for California planners is trying to define the term "normal." Currently, the prevailing wisdom in the state is that "California is a drought climate that is interrupted by storms. Droughts are not the anomalies; they are the norm" (Interviewee H 2022). In fact, a tree ring analysis shows century-long droughts in California's history, well beyond what the state has

DOI: 10.4324/9781003498537-4

faced in modern times (Lund et al. 2018). In light of these hurdles, California's unique water distribution and political battles have created a robust environmental culture, but one that is trying to claw its way out of the conflict and litigation that have plagued the state in the past.

Environmental Variables

Water Distribution and Infrastructure

California's geography is among the most diverse in the lower 48 states. The coastline features the only Mediterranean climate in North America, with coastal mountains that isolate this coastline from the interior of the state. According to NOAA, 80 percent of California's population lives within 30 miles of the coast, with most of the population on the arid southern coastline (NOAA Office of Coastal Management 2023). Moving inland from the coast, northern California features the Klamath and Cascade Mountains, between which begins the Central Valley of California. An average of 50 miles wide and 450 miles long, the Central Valley, also known just as "the Valley," parallels the Pacific Ocean and is the some of the most productive agricultural land in the world (CDFA 2018). The Valley is bordered in the east by the Sierra Nevada Mountains, the tallest mountains in the lower 48 states. Runoff and snowpack from the Cascades and Sierra account for the majority of freshwater in California, as most of the rivers in the state originate in these ranges. The eastern portion of the state lies in the rain shadow of the Sierra and consists of the Mojave, Colorado, and Great Basin deserts, the southern portion being called the "inland empire." This second agricultural region is primarily fueled by the southeastern border of California, the Colorado River.

California receives 75 percent of its rain and snow in the watersheds north of Sacramento; however, 80 percent of California's water demand comes from the southern two-thirds of the state (see Figure 4.1). Due to this diverse and rugged geography, water must be transported from the wet and snow covered north down to the Central Valley and dense population centers on the coast. In response to these challenges, the federal and state governments have combined to build one of the largest water networks in the world, the federal Central Valley Project (CVP) and the State Water Project (SWP). Originally built to provide water for the dry, but rich farmland in the Central Valley, the CVP has grown into a massive project including over 20 dams and reservoirs across 400 miles. In addition to the farmland, the CVP provides water to the greater Sacramento and San Francisco Bay areas, allocating over 9.5 million acre feet (AF) per year across all customers (USBR 2023).

Operating in conjunction with the CVP, the SWP is a series of reservoirs and pumping stations stretching across 700 miles and providing water to 27 million people and 750,000 acres of farmland (DWR 2019). While the CVP

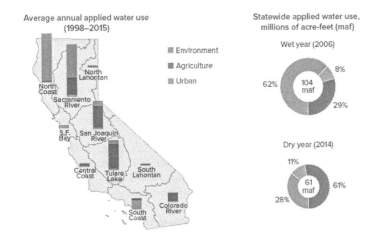

Figure 4.1 California's Annual Water Usage.

and SWP are engineering marvels, their water is over-allocated, with certain clients receiving their full allotment only four times in the past 30 years (Stern and Sheikh 2019).

From 1998 to 2015, the average water use was 50 percent environmental, 40 percent agricultural, and 10 percent urban, although the numbers fluctuated significantly depending on rainfall and snowpack. Environmental concerns in California have led to the prioritization of environmental water usage over storage and economic concerns. This is commonly referred to as "farmers vs. fish" (Bonatakis 2019). The percentage of water that is used for environmental versus agricultural uses varies greatly, as Figure 4.1 illustrates, but in dry years, the percentage of environmental withdrawals significantly decreases, placing strain on the rivers, estuaries, and wildlife supported by them.

One common sentiment is questioning what the natural environment in California would look like without the water redistribution. Water in California has been pumped from the delta region southward for almost a hundred years. Major bodies of water, such as Owens Lake and Lake Tulare, have been drained for farmland and urban water. The over-allocation and competing interests have led to legal battles between the two camps, followed by repeated appeals and deliberations (Boxall 2011; Arthur 2019). As described by one water expert,

The crux of the conflicts arises from the pinch point in the system—the Sacramento San Joaquin Rivers Delta. In order to deliver water to central and southern California, water must be pumped from the Delta and moved through a series of canals, reservoirs, and other facilities. Any time you build a large infrastructure such as the CVP and SWP, you alter the environment in a way that it can no longer function as an undisturbed, natural habitat. But we don't know to what extent. What would the habitat look like today without the CVP? We are operating to this hypothetical historic system from a hundred years ago that would likely would not exist with external factors like climate change.

(Interviewee R 2022)

The construction of infrastructure and subsequent redistribution of water have led to one of the most productive agricultural regions in the world. The United States is the top agricultural exporter in the world, and California is the top producing state. The farms of the Central Valley produce over one-half of the United States' fruits, vegetables, and nuts and over 20 percent of the total dairy production, which equates to a $50 billion per year industry (CDFA 2018). California's agriculture is responsible for more than 99 percent of total production in the United States for almost 20 key crops including almonds, garlic, and raisins. These key crops make up most of the produce aisle at grocery stores across the country. California is home to one of the five Mediterranean climates in the world, which means that it has a very long growing season and can grow almost anything (Interviewee R 2022).

Though most of the world relies on flood irrigation, many of California's farmers, including the largest agricultural district in the country, Westlands Water District, almost exclusively use underground irrigation systems. As described by an agricultural expert (Interviewee R 2022), "For over 90 percent of the water, once it enters the district, it doesn't see the light of day until it comes out as an almond or a tomato." This illustrates the significant investment the Westlands Water District has made into underground water distribution; it has effectively created one of the world's most advanced irrigation systems in the world based on underground sources. Moreover, this highlights the significant amount of water required to support agriculture in this region of the state.

While the agricultural production of California keeps the country's shelves stocked, within the state, agriculture's share of the state's GDP continues to decrease, accounting for approximately 3 percent of the state's GDP as of May 2022 (Interviewee R 2022). The tension between these two facts is at the heart of water politics in California. While the rest of the country relies on California's produce, agriculture is at a declining percentage of the state's revenue, with more and more money and influence flooding to the coasts, increasing the need for water in these historically arid regions.

Structural Variables

Rules and Communication

The rules and structures of the state water planning begin with the foundation of water law and then extend through the organizations that have developed to enforce it. Water in California is divided into two types, surface water and ground water. Surface water includes lakes, rivers, and reservoirs, while ground water is defined as water found in the spaces between sand, soil, and rock, known as an aquifer (DWR 2019). A mix of English water law and western geographic restrictions, California begins with the underlying principle that water used must be put to "beneficial use," which is codified in Article X, section 2 of the California Constitution (*Coffin v. Left Hand Ditch Co.*, 6 Colo. 443, 1882).

The structure of California surface law is a combination of riparian rights, appropriative rights, and prescriptive rights. Riparian rights are a hold-over from traditional English law that says that owners of land through which a stream passes have a right to its reasonable use, which the courts clarified to mean that riparian users may not deprive downstream users of their rights to the same waters (Persons 2017). Riparian owners are considered senior to other rights owners, except for certain appropriative rights existing from Spanish and Mexican laws that predate the California legal system.

Appropriative rights are independent of riparian rights and use the model of "first in time, first in right." Under this system, landowners that do not border streams and rivers can apply for a permit that allows them to divert surface water for the purpose of beneficial use (Persons 2017). As will be seen in the CVP, the first appropriative rights have priority over subsequent permits, establishing a strict hierarchy of water usage.

Finally, prescriptive rights were attained by unlawfully using another owner's rights for a minimum of five years, which could then be designated by a court as a legal right. According to the California Water Board (SWB 2019), these rights have not been available since 1914, as any new withdrawals must be permitted; however, many communities still maintain legal standing through prescriptive water rights (Persons 2017).

Groundwater is similarly subdivided into different rights, with overlying rights being most analogous to riparian rights, similarly followed by appropriative rights and prescriptive rights (Persons 2017). In California, groundwater has traditionally been unmanaged by the state, but as will be discussed later, the lack of surface water allocations during the 2012–2016 drought led to unsustainable withdrawals from the Central Valley Aquifer (Richey et al. 2015). As a result, recent legislation (such as Assembly Bill 1739 and Senate Bills 1168 and 1319) aims to create sustainability within California's groundwater, but does not go into full effect until 2042.

Because California's water is jointly managed by state and federal agencies, this legal relationship has particular salience. Traditionally, states were solely responsible for water management except for certain requirements, such as navigation along waterways. In the 1960s, this changed as the federal government took the lead on environmental policy, resulting in federal precedence over state enforcement. As deregulation took hold in the 1980s, the states once again took precedence over federal authority, leading to the current relationship where the federal government must apply for a state appropriation permit and comply with state regulations provided the regulations are reasonable (Abrams 2009). In the 1978 court case *California v. United States* (438 U.S. 645 (1978)), the Supreme Court ruled against the U.S. Bureau of Reclamation in a decision over water rights for the Melones Reservoir. The decision removed federal exemptions and required the federal government to comply with state environmental law. The decision removed federal exemptions from state regulations and established that the federal government can overrule state water allocations only with a "clear congressional command" (Abrams 2009, 405).

Water in California is controlled by the State Water Resources Control Board, commonly called the State Water Board (SWB), which operates as a child organization of the California Environmental Protection Agency (CalEPA). The five members are appointed by the governor to four-year terms and are responsible to ensure both quality and allocations of all water in the state (SWB 2019). Under the SWB, nine semi-autonomous regional water quality control boards are responsible for the quality control and local distribution of water within their regions (see Figure 4.2). These regional boards work in conjunction with the state to enforce state regulations and make recommendations for improving water quality and distribution to municipalities and special water districts (SWB 2019).

Separate from the water boards is the Department of Water Resources (DWR), which is responsible for managing all state-owned infrastructure, such as the California Water Project (DWR 2019). Just like any other consumer, the DWR must apply for water allocations from the SWB. While the regional boards have enforcement capabilities in the state, the majority of decisions and allocations are retained at the federal and state levels.

Most of California's surface water is managed through the federal Central Valley Project (CVP) and State Water Project (SWP). The Bureau of Reclamation within the U.S. Department of the Interior oversees the CVP on the federal level; however, allocations and coordination between the CVP and SWP are handled jointly between the state and federal agencies, a formalized relationship that dates back to the 1970s.

From a planning standpoint, water management and research are conducted by a myriad of organizations, on the local, regional, state, and federal levels. Unfortunately, as the California's flood plan manager stated, "Government entities often look at projects myopically . . . they often assume the project is good for the entire watershed, while only looking at a specific location.

REGIONAL WATER BOARDS

1	North Coast	6	Lahontan
2	San Francisco Bay	7	Colorado River Basin
3	Central Coast	8	Santa Ana
4	Los Angeles	9	San Diego
5	Central Valley		

Figure 4.2 California's Regional Water Boards.

There is no overall plan" (Interviewee T 2022). This disjointed management continues with the enforcement of ancillary programs, such as the Endangered Species Act (ESA). The ESA is managed on a species-by-species basis, with different species managed by different departments, often with contradictory outcomes. As an example, the 2008 and 2009 ESA biological opinions contradicted themselves and both could not be followed simultaneously (Interviewee R 2022). The lack of a holistic watershed or ecosystem management introduces conflict and confusion to the system.

Incentives

Incentives for water policy innovation vary by state and are often a result of grant opportunities and funding for water research. The state of California incentivizes water innovation and research, providing ample grant opportunities across the state. Grants are available for water research across a number of programs, from clean water to storage solutions. From 2014 to 2018, California gave an average of $1.11 billion of water-based grants to communities and organizations. The sizeable investment of funding for water programs and water research is evident when compared to other similar states, such as Texas's $172 million in grant funds in 2018 (TCEQ 2022).

Federal funding did not play a large role in the day-to-day water policy in California, but it did provide incentive and stability for large projects. The major water districts were relatively well funded for projects, and the state prioritizes environmental funding overall. The primary input for federal funding in California is the large-scale projects, such as the Delta tunnels or updates to the SWP. As stated by Interviewee M (2022), "Federal funding helps us build cooperation and get people to the table. Compensated projects are always better than uncompensated projects." So, while it does not have a large impact on policy selection it does allow important upgrades to existing infrastructure while encouraging cooperation among actors.

One of the ways in which California recently promoted water innovation is the facilitation of voluntary agreements between water users. The highlight of the Newsom administration has been the movement away from a pure regulatory model, in favor of agreement between all users in the system. Multiple interviewees discussed the myopic nature of water planning and the lack of holistic cooperation across an entire watershed (Interviewee T 2022; Interviewee R 2022). Similarly, the tenuous 49–51 agreements of the past only delayed the inevitable future conflict (Interviewee M 2022). The innovative solution for California is long-term negotiation involving all willing actors in the system, often taking multiple years to reach sustainable agreements.

Across all parties, from the farmers to the coastal regions, California water managers praised the use of voluntary agreements. As Interviewee R (2022) said, "Through voluntary agreements, we are able to address limiting factors that have not been addressed by regulations."

Largely based on Harvard University's "Getting to Yes," California has invested the resources and time to make voluntary agreements work. It is still too early to determine the longevity of these agreements, but the current trajectory is promising. While conflict is still a part of the process, there is hope that voluntary agreements can help stabilize the traditionally turbulent water process. This is not to say that the situation is perfect—there are still conflicting interests—but the state has decided to make this a priority and offer a roadmap for creative policy innovation in places that were once ruled by conflict.

Opening

Opening refers to the ability for citizen and interest groups to make their opinion known in state policy. Water in California is one of the most contentious government programs, with hundreds of opinions and inputs from organizations across the state. From 2011 to 2016, the SWB held an average of 21 board meetings per year, with many other workshops and committee meetings. Agendas are published well in advance with the opportunity for written comments, as well as for on-location comments and even remote meeting comments (SWB 2022).

Unlike in some of the other states, Californians take an active role in SWB meetings. During the 15 SWB meetings in 2011 alone, over 161 individuals commented on proposals. Ranging from private citizens to city officials and interest groups, public input to water board decisions is notably high, especially compared to other states in this study. The meeting minutes are detailed as far as the commenters and any presentations that were given to the board; however, during the span of this study, a word-for-word transcript was not available. Also, the SWB has removed many of its minute meetings during the drought portion of the study, although agendas and presentations are still available.

Regional and local water boards in California are equally open to public participation. As an example, both the Metropolitan Water District (MWD) in Southern California and the Westlands Water District (WWD) in the Central Valley conduct regularly scheduled open board meetings with periods of open comments for specific issues and policies (Interviewee M 2022; Interviewee R 2022).

While these entities pride themselves on being open to public comment, the actual decision-making processes are still often occluded. As the chief negotiator of a major water district described:

> One of the areas most challenged by water problems in the 21st century is the San Joaquin Valley. Today, the Valley faces the biggest water crisis in its history. It took a big step forward to address the crisis by creating the SJV Water Blueprint. The Blueprint was looking at innovative ways to solve their problems but made the serious mistake of excluding other interests from the process. As a result of the closed process, the term "Blueprint" was a four-letter word for the environmental NGOs and other interests because they didn't know what was going on in the process and didn't trust the farmers in the first place.
>
> Interviewee M (2022)

Interviewee M (2022) described the evolution of California's water culture as varying between adversarial conflict and cooperation. The early development era of California was almost entirely conflictual. As William Mulholland was taking water from across the state to feed the Los Angeles basin, he created enemies along the path, from farmers dynamiting aqueducts in Owens Valley to the governor of Arizona deploying his National Guard to prevent the Parker Dam on the Colorado River. This story continues up north, with the famed

John Muir fighting to prevent the city of San Francisco from damming the beloved Hetch Hetchy Valley. There was little opening in the development era of the state. Large cities took what they wanted and bullied those in their way (Interviewee M 2022).

By the 1980s, it became clear that this process was no longer sustainable, and some sort of regulation was required in order to protect the environment. This marked the transition to the regulation era (Interviewee M 2022). The regulation era was still largely conflictual and closed, but the variables that changed were the individuals who made the decisions and the desired outcomes. As Interviewee M (2022) described the rise of regulators, "They favored one objective and interest—the environment—over others and made one-sided decisions." Still ending in conflict, especially through litigation, the regulatory period simply traded one problem for another, resulting in new policy failures.

The 1990s saw the beginning of collaboration in California. This started in 1991 with the inclusion of best management processes, where entities were all brought together to negotiate the use of water throughout the complex state infrastructure (Interviewee M 2022). In 1994, the Monterey Amendment again brought the major water players together from across the state in a collaborative agreement on water management. This agreement "fundamentally changed expectations regarding the SWP and shifted responsibility for supply development in large part toward local agencies and away from the state" (Interviewee M 2022). That same year, the Delta Accord helped stabilize the San Francisco Bay Delta by continuing the collaboration that was began previously, across the state and federal agencies. While that agreement fell apart 11 years later, it helped stabilize the linchpin in California water, the Sacramento River Delta (Interviewee R 2022). This also demonstrated that cooperation was possible on a large scale, even if it was not lasting.

Inclusion and compromise have continued in certain forums throughout the state. The Water Forum Agreement in Northern California brought together water districts, local governments, businesses, and environmental groups for almost seven years' worth of negotiations. In the end, over 400 elected officials and environmental groups ratified this policy (Interviewee H 2022).

As Interviewee M (2022) described, the key to the victories of the 1990s was the inclusion of outside entities from the very beginning. As the chief negotiator for MWD, Interviewee M noticed that the agreements that were near-unanimous were the ones that stood the test of time. Seeking a 51 percent majority was bound to fail, because the 49 percent would always find a way to tear it down. "In the conflictual era, nothing was durable. It was 49–51 and those deals all fell apart." When discussing the success of collaboration, Interviewee M continued: "We can still find farmers who think the environmentalists are evil and water is 'wasted' flowing out to the ocean, but more and more their water managers aren't thinking that way. They are thinking: 'how can we sit down with these people and form networks?'"

The inclusion of environmental organizations into the state water process has been a net positive for the state. NGOs are often the experts in these topics, with at least ten that have direct input to the State Water Plan (Interviewee H 2022). This expertise and input were vital to negotiation. Interviewee M (2022) described the lessons learned from negotiating the continuous water deals throughout the state: "Any good collaboration threatens to fall apart every month or so. You have to leave your silo, hope the other side does the same, and make compromises to establish policy." Unfortunately, those compromises often upset the hardline idealists from each silo, creating resistance to collaboration.

Resistance

Resistance and conflict in environmental policy are a hallmark of the field, and in a state like California, litigation is a way of life for environmental actors. This phrase "farmers vs. fish" not only is a pithy headline but also truly encapsulates the feelings of many Californians (Bonatakis 2019). Almost every water policy in the state is going to end up in some form of litigation, even if the goal is to get a compromise in a completely different process (Interviewee R 2022).

From 2012 to 2016, over 4,500 "environmental" lawsuits were filed within the state of California, an average of two and a half lawsuits every day for five years (Justia 2022). The state of California's two primary water governance bodies, the State Water Resources Control Board and the Department of Water Resources, were directly involved in 439 of those lawsuits, while the Sierra Club, the most frequent interest group plaintiff, appeared in 192 of these cases. In fact, litigation is such an integral part of California Water that a federal judge in Fresno, Oliver Wanger, largely decided the state water policy for almost a decade, up to his retirement in 2011 (Boxall 2011). Unless there is near-unanimous agreement to policy, it is likely to end up in a court room.

While the landmark agreements in the 1990s helped usher in a period of collaboration, the persistence and reemergence of "winner takes all" groups have hampered some of the progress made in the past (Interviewee M 2022). As described by one water expert (Interviewee T 2022), "Too many people in water policy are trying to build their reputations." Cooperation is a long-term solution, often extending beyond the career of an individual. The desire for a short-term win often overwhelms the benefit of long-term cooperation (Interviewee M 2022; Interviewee T 2022).

One point highlighted by multiple interviewees is the inability of regulators to address holistic solutions. The primary lever for regulators in California is managing water flows. As stated by Interviewee M (2022), "regulators don't have the tools to solve these complex water problems. They have one tool—flow—and flow alone can't solve these problems." Interviewee R (2022) echoed these sentiments, saying that "previous cuts to the water have not reversed the losses in the ecosystem, but it's the easiest lever to pull, so it is the most used." Even though previous cuts did not restore the ecosystems, regulators have continued to cut almost half of the water to the southern Central Valley, with no appreciable environmental improvement (Interviewee R 2022).

Another source of conflict in the state is the physical distribution of water and disproportionate influence of particular districts. As an example, the Westlands Water District is the largest agricultural district in the country, comprising over 1,000 square miles along Interstate 5 in the western portion of the Central Valley, and supporting over $5 billion of economic activity annually (Interviewee R 2022). Despite the massive economic impact of the district, the relatively small population results in only two state legislators in the region. Conversely, the Metropolitan Water District of Southern California covers 185 miles of densely packed cities along the coast, servicing over 19 million customers from Oxnard to San Diego. As a result of this massive population and resultant political clout, the MWD is often referred to as the "800-pound gorilla" in California water, since the district includes over half of the California Legislature and Senate seats (Interviewee M 2022; MWD 2022). When these politically powerful districts are willing to negotiate and compromise, progress can occur. But, since the big districts have abandoned collaboration in the past decade or so, "it hasn't worked out well for them" (Interviewee M 2022).

Finally, the state has intentionally established restrictions on outside inputs to water policies. During the Trump administration, the Bureau of Reclamation proposed raising the dam at Lake Shasta by 18.5 feet to provide more storage and water stability for the state. At a cost of $1.4 billion, the federal government proposed paying half of the share, with the other half coming from the state of California, or other parties. The Westlands Water District, outside of Fresno, began an environmental review that would answer the question of whether the project would adversely affect the trout fishery or the free-flowing state of the river. As the largest irrigation district in the country, Westlands stands to benefit greatly from the increased storage of Lake Shasta with more annual water allocations over the current agreements. The attorney general of California sued Westlands, since the California Wild and Scenic Rivers Act (CA 5093.50–5093.71) outlines that only the Department of Water Resources can assist federal, state, or local environmental studies into the expansion of Lake Shasta. As the Westlands spokesperson stated, "We are not sure expanding the dam would have an adverse effect on the river, but we were sued from even investigating the question, so we decided not to pursue" (Interviewee R 2022).

To be sure, the expansion of Lake Shasta was as much of a political tool, given the conservative voters and representatives in the Central Valley, as it was an environmental tool to stabilize water quantities. This case further highlights the use of litigation in California, where the state will sue one of its own regional water entities. The attorney general of California has frequently

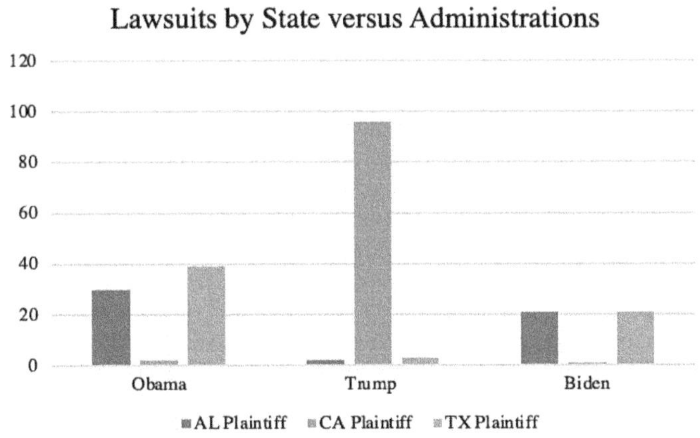

Lawsuits by State versus Administrations

Figure 4.3 State Lawsuits Against Presidential Administrations.

sued the federal government over policy positions (see Figure 4.3). Of the 135 lawsuits against the Trump administration, California was a plaintiff in 94 of them, and initiated 52 of the lawsuits, second only to New York's 64 (Nolette 2022). When California wants to speak up, litigation is the primary tool.

Cultural Variables

Shocks

Major shocks to the states and citizens are the spark that drives innovation in policy, and shocks in California water are nothing new. In 1861–1862, the area around Sacramento experienced a megaflood that left approximately 30 feet of water over some areas for weeks after it started (Huang and Swain 2022). On the other hand, catastrophic droughts are equally present in the psyche of Californians, with repeated cycles of too much and too little water.

In Southern California, MWD conducted a reliability study in 1986 that showed that with the population growth of the 1980s, the district had a 50 percent reliability from year to year, meaning that in any given year, the district would likely not have enough water to sustain the residents' requirements (Interviewee M 2022). The very next year, a five-year drought began that tested their newly found weakness. The SWP, which was the primary source of water to Southern California, was built to stabilize the region using the 1928–1934 drought model. The 1987 drought was very similar to the model on which the SWP was built, and as Interviewee M described, the SWP "failed miserably." One of the painful lessons learned from the drought was that "Metropolitan

Water can't rely on the SWP. We can't rely on imported water" (Interviewee M 2022).

California's drought from 2012 to 2016 was caused by a combination of unusually low rain and snowfall and unusually high temperatures. These four years were estimated as a "once in every 1,200 years" drought, although longer and drier periods exist in the climate record (Lund et al. 2018). As the drought wore on, surface water storage dwindled, leaving many reservoirs dry and causing water contractors, such as the Friant Water Authority, to receive zero water deliveries for over two years (Lund et al. 2018).

Even before the drought, groundwater was used extensively throughout the Central Valley, from farmers to municipal water districts. As the southern region of the Central Valley began to feel the impact of water restrictions, many farmers expanded their use of groundwater. In addition to the economic impact, as aquifers are drained, water quality decreases until water is technically available, but unusable due to mineral concentrations. Over the span of the drought, it is estimated that Californians pumped 10 cubic miles of groundwater, equivalent to about one-third of the total volume of Lake Tahoe (Friedlander 2018). Due to this unsustainable pumping, the ground subsidence in the region increased almost two feet over the span of the drought (Murray and Lohman 2018). In fact, Murray and Lohman (2018) found that between Spring 2011 and Spring 2017 approximately 86% of wells in the Tulare Lake region reported groundwater levels 1.5 meters lower as compared to Spring 2011. The combination of drought conditions and unsustainable withdrawals make the Central Valley Aquifer the most stressed aquifer in the world (Richey et al. 2015).

Environmentally, the drought killed years' worth of salmon hatcheries, causing fear at the time that the Chinook salmon might not recover (Bland 2015); however, recent evidence shows that the fish populations may have been more resilient than initially believed (Chea 2019). Additionally, an estimated 102 million trees died in California's forests, fueling subsequent wildfires that crippled the state (Lund et al. 2018).

The economic impact of water reduction was equally staggering. It is assessed that in 2015, agricultural losses from the drought totaled $2.7 billion, with an estimated 21,000 direct and indirect job losses, primarily located in the Central Valley (Howitt et al. 2015). Finally, the subsidence rates have caused untold economic and environmental damage, such as damage to roads and pipes, in addition to bowing in the California Aqueduct, which is restricting the amount of water that can be delivered to southern California (Friedlander 2018).

One of the biggest shocks for Californians was the lack of clean drinking water within communities in the Central Valley. As Interviewee H (2022) stated:

Contamination of ground water left many communities behind, primarily in the Central Valley, which had no real political clout. "Water for People" has very real face in places like East Porterville, CA. . . . Mass population centers on the coast don't even know where these communities are, but a few activist groups kept the pressure on the California Legislature. Seeing people walk miles for water shocked the legislature, which helped lay the foundation for SGMA and managing our groundwater resources.

The geographic and cultural disconnect between decision-makers living on the coast and farming communities cannot be overstated. California is a large, geographically diverse state, with the vast majority of the population, including most lawmakers, living in a relatively small region along the coast, far removed from the impoverished, water-starved communities that provide the stability to the coast.

Framing

One of the most difficult issues with environmental policy is defining the problem and framing it in a way that garners attention. Each side is hoping for a solution that resembles their desires, often in sharp contrast to other groups. To be clear, there is a finite but variable amount of water in California. Water is closely monitored, and potential solutions are often a zero-sum game. As such, how the problem is framed has a direct impact on policy outcomes, with water being shifted from one group to the next.

For the Sierra Club, many of California's water problems center around the diversion of water from the Delta toward other users, which has an environmental impact on species throughout California. According to its 2013 report, diversion of water from the Delta toward the south has "collapsed" the ecosystem (Sierra Club California Water Committee 2013). Prior to the period of this study, from 2008 to 2010, the amount of salmon migrating back into the Northern California rivers was so low that the state canceled the commercial salmon seasons. Other species saw similar declines in populations. Interestingly though, the Sierra Club did not mention that just prior to its 2013 report, the 2012 salmon season saw the largest salmon season off the coast of Oregon since the measurements began in 1985 (Kinney 2012).

In addition to water conservation, the Sierra Club's focus is on maintaining the current environmental status. Solutions of the past, such as damming rivers and creating storage in reservoirs are unfortunate realities for groups like the Sierra Club; however, they are committed to not expanding these projects, and problems are carefully defined as environmental problems that are the result of human actions.

In contrast to environmental groups, farming groups in the Central Valley, such as the Westlands Water District, have defined the problem as a mismanagement of California's water resources based on bureaucratic hurdles and poorly defined science, which has a very real economic impact and human toll in underserved communities. As Thomas Birmingham, the general manager of Westlands stated, "The needs of our community and the livelihood of our workers are being sacrificed due to questionable decisions by federal officials to protect Delta smelt" (Westlands Water District 2014).

For farming regions such as the Central Valley, the water problem is not just defined economically. Multiple communities, such as the city of Huron south of Fresno, saw their wells running dry in the heart of the drought. After receiving zero water allocation for multiple years and appeals for federal aid denied by the Bureau of Reclamation, Huron reached out to Westlands to stabilize its water supplies (Westlands Water District 2014). Other towns, such as Avenal and East Porterville, saw similar stories of taps running dry, which helped define the human impact of the problem for lawmakers who only drive through the Central Valley on their way to Sacramento from the coast.

While environmental groups and farming groups each frame the problem from opposite extremes, the state has been careful to remain neutral when possible. In Governor Brown's 2014 State of the State address, he highlighted both sides of the problem, from safe drinking water in disadvantaged towns in the Valley, to wetlands restoration. Governor Newsom built upon that legacy by promoting the voluntary agreement process, trying to define the problem and potential solutions from multiple angles, with overwhelming success thus far (Interviewee M 2022; Interviewee R 2022).

Each of these groups has a different desired outcome. Environmental groups want more water in the Delta and frame the problem in terms of the human impact upon native species. Farming groups want more water flowing south and frame the problem as the economic and basic livelihood of communities in the Valley, which are sacrificed for negligible environmental improvement. The state tries to play a middle road with coequal interests, while enabling cooperative outcomes to which each side can compromise. Collectively, they are all appealing to different people, with little overlap in constituencies.

Increased Agenda Attention

As the problems are framed, the increased public attention should be measured in both news outlets and public interest. Throughout the period of study, regardless of the location, Californians remained acutely aware of droughts in other states (Figure 4.4). This corresponds to research illustrating how topics that gained the public attention, but subsequently faded, are more likely to recapture attention when they reemerge (Downs 1972). The 2011 drought in Texas, though a thousand miles away, captured the average Californian's attention.

The link between media and public attention appears to be more complex than a direct causal relationship. The media certainly plays a role in public consciousness, but that correlation appears to be different based on the proximity to the story. The Texas drought in 2011 held the national attention for

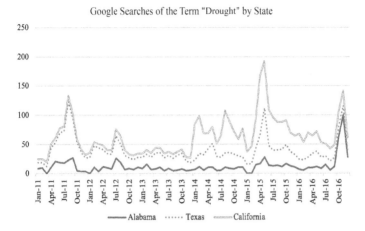

Figure 4.4 Google Searches of the Term "Drought" by State.

almost a year, which corresponded to the California attentiveness. However, within the state, major events correlated with a spike in attention in real time before the traditional news cycle was able to report. In this sense, traditional media is certainly a vehicle for public consciousness, but it is not the only one. Other variables may have more of an impact on local consciousness than traditional media, such as personal experience, Twitter, and Facebook.

Furthermore, large rain events, such as Hurricane Norbert, which dumped 3 inches of rain in southern California, and 6 inches of rain in parts of Arizona, immediately correlated to a Google Trends (2022) spike questioning, "Is the drought over?" This did not correlate to a spike in *Los Angeles Times* (2022) stories mentioning drought, showing that the news outlets were not correlating a big rain event to drought status, unlike the average population (Figure 4.5).

Legitimacy

How states and interest groups attempt to legitimize their proposed solutions to problems is indicative of their view of the problem, and the potential implications of those solutions. The water issue in California has largely been reduced to defining what constitutes a beneficial use of water and what is a wasteful use of water.

The state of California is careful to walk a tightrope between the multiple interests in the state. California policy expresses "coequal goals," where water for the environment is just as important as water for the people (Interviewee H 2022). The DWR website (2022) discusses the careful balance between both human and environmental needs for water. DWR highlights the benefit of agriculture, while balancing the need to be responsible: "California's

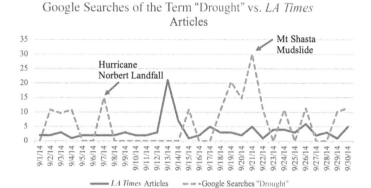

Google Searches of the Term "Drought" vs. *LA Times* Articles

Figure 4.5 Google Searches of the Term "Drought" Versus *Los Angeles Times* Articles.

agricultural success can be attributed to irrigation, the act of supplementing rainfall during times of no precipitation to meet the water demands of plants . . . applying the right amount of water at the right time not only ensures agricultural crops' growth, but also aids in conserving the state's limited water supply." Without water to the ocean, sea water intrusion will salinate the groundwater, ruining the riparian habitat. Planning for the state also assumes a 10 percent decrease in future water availability, requiring a restructuring of current water agreements.

Much of the focus of state messaging geared toward residential water. Although it is only about 10 percent of the state's water consumption, the push toward more climate-native lawns in urban areas has been quite effective. As the state's Save Our Water website (DWR 2022) states, "One of the most effective ways to do this [reduce consumption] is by removing or reducing the size of your lawn. About half of residential water use is spent outdoors on lawn and landscaping. . . . Your yard will require very little maintenance and care, you'll promote a healthy ecosystem and improve the health of your soil, and you'll never have to mow again." The DWR website (2022) even has coloring sheets for kindergarten classes and above, with tips on how to conserve water, as well as to remind their parents when they are wasting water. While this messaging from the state has been effective, with a 7.5 percent reduction in urban water use from June 2020 to June 2022, this is still a relatively small percentage of the state water usage (Newsom 2022).

From the farmers' perspective, Westlands Water District's 2017 water plan describes the diversion of water in support of the Chinook Salmon and Delta smelt as "an unfair and inequitable burden on those water users south of the Delta" (WWD 2017, 12), while highlighting the economic impact of court-ordered reductions on farm workers and communities. Westlands points out that the system has been altered to the point of unrecognizability from the native Delta over a hundred years ago. In fact, Westlands argues that no one can know what that unaltered environment would look like today, especially given climate change the shifting populations (Interviewee R 2022). Proposed solutions for Westlands are largely focused on increasing overall water flow, from the California Delta Tunnels, which would move water around the Delta, to increasing storage in Lake Shasta.

Conversely, the Sierra Club California (2022) opposes any change to storage or moving water from the Bay Delta to the south. Repeatedly, press releases (July 2012; May 2013; December 2013; June 2014; April 2015; December 2015; February 2016) condemn the expansion of water storage, building tunnels for shipping water from the Delta, and groundwater storage. The Sierra Club (2016) states that California needs "fresh thinking" while the state's drought plans are based on "old ideas that simply continue an unsustainable water system." Ideas proposed by the Sierra Club include recycled water, increased conservation and efficiency, as well as groundwater management for urban and agricultural uses. Additionally, they oppose ocean desalination due to the expense of the projects and environmental impact of brine.

Potential solutions for environmental groups clearly focus on human activity vice storage. Many Californians are still uneasy with the expansionist past and the requirements to provide services to such a large population. With this understanding, environmental groups like the Sierra Club want to curb behavior, not expand reservoirs, which puts it clearly in opposition to other groups in the state.

Each side has valid points, but also valid concerns. Acres of water-hungry plants, such as almonds, seem greedy in a drought-prone region, while watching underserved residents walk miles because they didn't receive any water allocations for the past three years seems inhumane. The push toward voluntary agreements will hopefully provide stability moving forward; however, the track record of agreements makes that statement cautiously optimistic.

Policy Innovations (2011–2016)

The drought from 2014 to 2016 pushed California to the breaking point, which resulted in multiple innovations. California's first drought responses were focused on urban usage, to include a 20 percent voluntary reduction in use, limited lawn watering, hotel water restrictions, and restaurants serving water only on request. These measures failed to reach the 25 percent overall reduction in water usage, causing the state to take more restrictive measures (Persons 2017).

Due to dramatic drops in groundwater levels, in September 2014 California passed a three-law package (Calif. AB 1739, SB 1168, SB 1319), collectively known as the Sustainable Groundwater Management Act (SGMA, pronounced Sigma). Intended to create sustainable groundwater, these laws require local water boards and communities achieve sustainability no later than 2042 (DWR 2019), a prospect particularly worrisome for many communities that rely exclusively on groundwater for public utilities, especially those who are last in line for surface water allocations.

SGMA has revolutionized how public utilities view groundwater. While there are still unsustainable withdrawals due to the long timeline for full implementation, regions have created Groundwater Sustainability Agencies (GSAs) that are responsible to the DWR for groundwater management, subsidence, and water quality (South Fork Kings Groundwater Sustainability Agency 2022). The accountability on a local and regional level is a game-changer for California groundwater and presents a pathway for success in the future (Interviewee T 2022).

As previously mentioned, the other major innovation is the use of voluntary agreements with the water districts throughout the state (Blumenfeld and Crowfoot 2019). Voluntary agreements have brought actors to the table, encouraging cooperation in the place of litigation. While it is too early to judge the longevity of these agreements, every interviewee described them as a major policy innovation within the state. The willingness to invest years of time in long-term agreements shows not only the importance of this process to the state but also the enthusiasm of these actors to solve the water issues. It is easy to walk away from the table. It is difficult to spend seven years on a negotiation, but that is what this process has enabled, and the early results have been positive.

Conclusion

This research was built around four propositional statements. This conclusion will evaluate each proposition based upon the findings from this case.

(1) States with highly detailed, frequently reviewed water plans (<ten years) will be less likely to enact non-incremental policy innovations in response to a new drought event.

California's water planning is an effective, top-down water planning system. The state has effective monitoring over every drop of water within its borders, with diversified sources and in-depth planning. With that said, the state has over-allocated its water resources, with a full delivery of water on average of one out of ten years. Although the state takes great pride it its water management, water districts in the southern Central Valley routinely face minimal water allotments, even receiving zero surface water for multiple years.

This inequitable water system led to unsustainable groundwater withdrawals that pushed the environment to the brink in the Valley. The lack of previous groundwater management actually allows for a multifaceted test of this proposition. Since surface water is highly planned and frequently reviewed, innovations in surface water were small incremental improvements. However, the complete lack of management of groundwater led to sweeping, non-incremental innovations following the 2014–2016 drought. In this sense, California provides positive support for this proposition as both incremental and non-incremental innovations.

(2) States with distributive water infrastructure will experience a more competitive and fractured water political culture.

California's traditional water culture is the very definition of competitive and fractured. Litigation has been the status quo for California for decades. One of the primary reasons for this is the fact that a drop of rain up by Mt. Shasta along the Oregon border can be distributed anywhere from San Francisco to San Diego, with hundreds of miles of farmers along the way. As previously stated, the water in California is over-allocated, so water is a zero-sum game in the state. The ability to transfer water hundreds of miles to millions of people is the great blessing of California, but it is also the curse. While the state has made great strides in voluntary agreements, the longevity of these agreements is yet to be tested. As such, currently this research supports his proposition, although future analysis of long-term stability could alter these findings.

(3) States with competitive environmental cultures will have less innovative policies.

California has managed to have both a highly competitive environmental culture and innovative policies. One of the reasons for this, somewhat outside of this research, is that the state is effectively a one-party state, with the large farming industry represented by a very small minority in the state legislature. As such, policy on the state level is relatively stable, with lurches largely the result of judicial inputs. Due to these contradictions in culture and policy, this case study does not support this hypothesis.

(4) States whose citizens have experienced major water shocks will have a more cohesive, innovative policy culture.

The average Californian is very attuned to water conservation. Despite the conflicts between regions in the state, Californians have a strong connection to their environment and a deep understanding of the dangers of both drought and floods. The major droughts of the past have left an indelible mark on the Californian culture, even if the effects of the drought are largely insulated from the average resident. Citizens largely do their part to contribute to water savings, although urban use is a small portion of water allocation. The impact of conflict in California is felt most among the extreme interests, from environmentalists to farmers. The one major exception to this is the small communities on the fringe of the Central Valley. These citizens feel the droughts much more deeply than the average Californian on the coast. This case study strongly supports this propositional statement. Californians have a strong environmental culture, largely based on the shocks and impacts of droughts in the state.

Overall, California is one of the core case studies for water innovation. Although idiosyncratic in its culture and infrastructure, the state provides great insight to the impact of the environment on water policy. From interest groups to citizen impact, California experiences the full spectrum of environmental policy and acts as a model for other states to emulate.

References

Abrams, Robert H. 2009. "Water Federalism and the Army Corps of Engineers' Role in Eastern States Water Allocation." *University of Arkansas at Little Rock Law Review* 31 (3): 395. Accessed October 31, 2019.

Arthur, Damon. 2019. *After AG sues, Westlands Water District Says it's Studying Whether to Support Shasta Dam Raise,* May 14. https://www.redding.com/story/news/2019/05/14/state-ag-environmental-groups-sue-stop-shasta-dam-raise/3668909002/

Bland, Alastair. 2015. *For California Salmon, Drought And Warm Water Mean Trouble.* January 5. https://e360.yale.edu/features/for_california_salmon_drought_and_warm_water_mean_trouble.

Blumenfeld, Jared, and Wade Crowfoot. 2019. "Voluntary Agreements Progress Report." *California Natural Resources Agency.* July 1. http://resources.ca.gov/wp-content/uploads/2019/07/Voluntary-Agreement-Progress-Report-07-01-19.pdf.

Bonatakis, Lauren. 2019. *Farmers vs. Fish: The Story of Delta Smelt.* June 24. https://envirobites.org/2019/06/24/farmers-vs-fish-the-story-of-delta-smelt/.

Boxall, Bettina. 2011. *The Man With His Hand on California's Spigot.* October 7. www.latimes.com/local/la-xpm-2011-oct-07-la-me-water-judge-20111007-story.html.

California Department of Food and Agriculture (CDFA). 2018. *California Agricultural Production Statistics.* www.cdfa.ca.gov/statistics/PDFs/2017-18AgReport.pdf.

California Department of Water Resources (DWR). 2019. *Water Storage and Supply.* Accessed 12 February, 2019. https://water.ca.gov/What-We-Do/Water-Storage-And-Supply.

California Department of Water Resources (DWR). 2022. *Save Our Water—Removing Your Lawn.* May 17. Accessed October 6, 2022. https://saveourwater.com/en/News-and-Events/Latest-News/Removing-Your-Lawn.

California State Water Board (SWB). 2019. *Water Boards' Structure.* www.waterboards.ca.gov/about_us/water_boards_structure/.

California State Water Board (SWB). 2022. *Upcoming Events*. Accessed November 12, 2023. www.waterboards.ca.gov

California v. United States (438 U.S. 645 (1978).

Chea, Terrence. 2019. *Off the Hook: California King Salmon Rebounds after Drought*. August 22. www.businessinsider.com/off-the-hook-california-king-salmon-rebounds-after-drought-2019-8.

Coffin v. Left Hand Ditch Co., 6 Colo. 443, 1882.

Cohen, Michael P. 1988. *The History of the Sierra Club, 1892–1970*. San Francisco: Sierra Club Books.

Dettinger, Michael. 2011. "Climate Change, Atmospheric Rivers, and Floods in California–A Multimodel Analysis of Storm Frequency and Magnitude Changes." *JAWRA Journal of the American Water Resources Association* 47 (3), 514–523. https://doi.org/10.1111/j.1752-1688.2011.00546.x.

Downs, Anthony. 1972. "Up and Down with Ecology: The Issue-attention Cycle." *The Public* 28: 38–50.

Friedlander, Blaine. 2018. *Groundwater Loss Prompts More California Land Sinking*. August 29. https://news.cornell.edu/stories/2018/08/groundwater-loss-prompts-more-california-land-sinking.

Google Trends. 2022. Accessed November 12, 2023. https://trends.google.com/trends/

Howitt, Richard, Duncan MacEwan, Josue Medellin-Azuara, Jay Lund, and Daniel Sumner. 2015. *Economic Analysis of the 2015 Drought for California Agriculture*. Davis, CA: UC Davis Center for Watershed Sciences.

Huang, Xingying, and Daniel L. Swain. 2022. "Climate Change Is Increasing the Risk of a California Megaflood." *Science Advances* 8: eabq0995. https://www.science.org/doi/pdf/10.1126/sciadv.abq0995

Interviewee H in discussion with the authors. May 20, 2022.

Interviewee M in discussion with the authors. August 4, 2022.

Interviewee R in discussion with the authors. August 31, 2022.

Interviewee T in discussion with the authors. July 27, 2022.

Justia. 2022. *Justia Dockets and Filings*. Accessed September 5, 2022. https://dockets.justia.com/.

Kinney, Aaron. 2012. *California Gets Longest Commercial Salmon Fishing Season Since 2005*. April 5. Accessed October 30, 2022. www.mercurynews.com/2012/04/05/california-gets-longest-commercial-salmon-fishing-season-since-2005/.

Lund, Jay, Josue Medellin-Azuara, John Durand, and Kathleen Stone. 2018. "Lessons From California's 2012–2016 Drought." *Journal of Water Resources Planning and Management* 144 (10): 04018067.

Los Angeles Times. 2022. *LA Times Archives*. Accessed October 10, 2022. www.latimes.com/archives.

Metropolitan Water District of Southern California (MWD). 2022. *Fact Sheets and Publications*. Accessed November 12, 2023. www.mwdh2o.com/fact-sheets-and-publications/

Miller, Kenneth. 2020. *Texas vs. California: A History of Their Struggle for the Future of America*. New York: Oxford University Press.

Murray, Kyle D., and Rowena B. Lohman. 2018. "Short-lived pause in Central California subsidence after heavy winter precipitation of 2017." *Science Advances* 4 (8): eaar8144.

NOAA Office for Coastal Management. 2023. "Understanding and Planning for Sea Level Rise in California." https://coast.noaa.gov/digitalcoast/stories/ca-slr.html.

Newsom, Gavin. 2022. "Governor Newsom Announces Water Strategy For a Hotter, Drier California." August 11. Accessed November 12, 2023. www.gov.ca.gov/2022/08/11/governor-newsom-announces-water-strategy-for-a-hotter-drier-california/

Nolette, Paul. 2022. "State Litigation and AG Activity Database." Accessed November 12, 2023. https://attorneysgeneral.org.

Persons, Bonnie. 2017. *Water Shortage and Water Law: The Impending Crisis in Semi-Arid Climates*. Accessed October 31, 2019. *Journal of Comparative Urban Law and Policy* 2(1): 1–43.

Richey, Alexandra S., Brian F. Thomas, Lo Min-Hui, John T. Reager, James S. Famiglietti, Katalyn Voss, Sean Swanson, and Matthew Rodell. 2015. "Quantifying renewable groundwater stress with GRACE." *Water Resources Research* 51: 5217–5238.

Sierra Club California. 2016. "Senator Feinstein's Water Bill Seriously Flawed." February 10. Accessed October 30, 2022. www.sierraclub.org/sites/default/files/sce/sierra-club-california/PDFs/Senate%20Water%20Bill%20Flawed_Sierra%20Club%20Cal_2.10.16.pdf

Sierra Club of California. 2022. *Press Releases*. Accessed October 30. www.sierraclub.org/california/press-releases.

Sierra Club California Water Committee. 2013. "Clean, Sustainable and Reliable Water Supply: Alternatives to the Giant Bay Delta Tunnels." *Sierra Club of California*. Accessed October 30, 2022. www.sierraclub.org/sites/default/files/sce/sierra-club-california/PDFs/Sierra percent20Club percent20CA percent20Alternatives percent20to percent20the percent20Tunnels percent20Format percent20Revised.pdf.

South Fork Kings Groundwater Sustainability Agency (South Forks GSA). 2022. *South Fork Kings GSA*. Accessed November 2, 2022. https://southforkkings.org/.

Stern, Charles V., and Pervaze A. Sheikh. 2019. *Central Valley Project: Issues and Legislation*. Washington, DC: Congressional Research Service.

Texas Commission on Environmental Quality (TCEQ). 2022. "Texas Commission On Environmental Quality Expenditures by Object of Expense For the 4th Quarter Ended August 31, 2018." Accessed October 10, 2022. www.tceq.texas.gov/downloads/agency/administrative/expenditures/fy18–4th-qtr-expenditures.pdf.

United States Bureau of Reclamation (USBR). 2023. "Central Valley Project Water Quantities for Delivery." Accessed November 26, 2023. www.usbr.gov/mp/cvp-water/docs/cvp-water-quantities-for-delivery-2023.pdf.

Westlands Water District (WWD). 2014. "Statement by Thomas Birmingham upon Passage of the Emergency Drought Relief Act." *WWD Press Releases*. May 22. Accessed November 1, 2022. https://wwd.ca.gov/wp-content/uploads/2014/05/emergency-drought-relief-act-passage.pdf.

Westlands Water District (WWD). 2017. "Westlands Water District Water Management Plan 2017 Criteria." Accessed October 10, 2022. https://wwd.ca.gov/wp-content/uploads/2022/11/water-management-plan-2017.pdf.

5   Alabama

Benign Neglect

The state of Alabama is one of the most water-rich states in the United States. Located in the southeastern United States, Alabama begins with the foothills of the Appalachians along the northeastern third of the state and slowly returns to sea level along the Gulf Coast. The land is heavily wooded and crossed by rivers primarily flowing from the northeast to the south by southwest. Alabama's geography has enabled an incredible biodiversity, from amphibians to fish and trees, it is considered as one of the most biodiverse states in the country (Jenkins et al. 2015). The state, as well as many of its residents, takes great pride in its natural water resources, even illustrating and featuring its abundant river systems in the official state seal. Despite multiple droughts in the past decade, it is the only state in the region that does not have a comprehensive water plan, instead opting to push water planning to the local level (ADECA 2018). It does, however, have a statewide Drought Management Plan in fulfillment of the Alabama Drought Planning and Response Act of 2018.

Alabama is one of the most politically conservative states in the country. Pew Research Center (n.d.) ranked Alabama as having the highest percentage of conservatives (50%) and the lowest percentage of liberals (12%) in the United States. This ideology shapes policy outcomes relative to water as there is generally a pro-business and growth, limited government, and limited regulatory intervention mentality among policymakers (Pew n.d.). However, there is a unique opportunity for environmentalists. The Deep South is also an outdoors haven, with high percentage of sportsmen and women. In fact, over 13 percent of all Alabamians pay for an annual hunting license, so while many may be politically conservative, many Alabamians are also concerned about maintaining a stable and pristine environment for recreation (Mueller 2015). The combination of these factors—water abundance, pro-business views, and outdoors culture—creates a unique water environment but one that has yet to produce a state water plan. While this is certainly influenced by variables such as political ideology, there are other factors as well. Politically liberal states, such as Washington and Connecticut also do not have state water plans, while politically conservative states like Texas, Montana, and Idaho have recurring

DOI: 10.4324/9781003498537-5

plans. This hints at other variables that combine to facilitate or inhibit innovative policy outputs/outcomes.

Prior to examining innovation (or the lack thereof) within Alabama's water management/drought mitigation strategies, it is helpful to provide a cursory snapshot of Alabama's water "environment," as noted in the following:

- About 44 percent of the state's population relies on groundwater for drinking water, whereas approximately 56 percent of the population depends on surface-level resources;
- The state's annual average surface water supply is about 33.5 trillion gallons; this water traverses across the state's 14 river basins and coastal drainage regions. Of this amount, slightly less than 20 trillion gallon is the result of annual rainfall runoff;
- There are more than 132,000 miles of rivers and stream channels in Alabama, yet the state is home to only one natural lake;
- Alabama's groundwater supply is approximately 550–555 trillion gallons of freshwater, which is spread among 19 major aquifers or aquifer systems across the state.

(ADPH 2023; Hairston et al. 2023)

Despite droughts in the state and the promulgation of a 2018 drought management plan, the political will to invest resources in overall water planning has not overcome the opposing business interests and state financial priorities. One point of interest in Alabama is the purpose of water in the state. While this research is primarily looking at drought response, many Alabama water experts framed water planning in the context of economic development. Alabama was once an agricultural powerhouse, especially in the "black belt," the stretch of fertile soil extending from south of Demopolis, Montgomery, and Tuskegee, to Auburn. Since the collapse of the agrarian economy in the early 1900s, this region has languished economically outside of the major cities in the state (Interviewee B 2022). Whereas most states are driven to water management by the fear of supply issues, Alabama explicitly links water with economic prosperity, especially for regions that have struggled to regain their economic strength (Interviewee B 2022).

Prior to the early 1970s, Alabama was engaged in almost no water resources planning. The passage of the 1965 Water Quality Act, which included new requirements for states to develop ambient standards for water quality within their borders for all interstate waters (Morris 2022, 68), prompted a discussion around water resources in the state, although most of the discussions around water fell under the purview of economic development entities in Alabama. The Federal Water Pollution Control Act of 1972 further solidified the requirements for ambient standards and extended the requirements to intrastate waters as well (Morris 2022). The state responded with two reports published in the early 1970s that made several recommendations around water resources planning. While these recommendations discussed the need for ensuring an adequate water supply, they did little to address drought resilience even though Alabama had endured several droughts during the 20th century.

A second effort to address water planning began in 1989 with the formation of the Water Resources Study Commission, created by the executive order of the state's governor, H. Guy Hunt, the state's first Republican governor since Reconstruction. Four years later, the state legislature passed the Alabama Water Resources Act of 1993, largely in response to drought conditions in the 1980s and ongoing water rights conflicts with Georgia and Florida in the Chattahoochee River basin. Among other issues, the legislation created an office designated to become the lead actor for drought management, water rights conflicts, and floodplain management. Notably, the office was located in the Alabama Department of Economic and Community Affairs, solidifying water supply as an economic issue in the state (Andreen 2016).

A third effort to address water supply began in 2007, a period of exceptional drought in much of the state. In this round of activity, the legislature created a joint legislative committee (the Permanent Joint Legislative Committee for Water Policy and Management) to address water management and policy issues in the state. Legislation followed a year later that recommended several policy actions related to drought management, but notably the legislature declined to appropriate funds to the tasks. In 2011 the Alabama Water Agencies Working Group was created, an interagency group to bring together different stakeholders and develop a statewide plan (Alabama Water Agencies Working Group 2012). This eventually led to the passage of the Alabama Drought Planning and Response Act in 2014. This legislation stands as the most recent effort by state government to address issues of drought.

Environmental Variables

Water Distribution and Infrastructure

The state of Alabama is home to 132k miles of rivers and streams, with 563k acres of ponds, lakes, and reservoirs (Alabama Rivers Alliance 2023a). Alabama receives the third most annual rainfall in the continental United States at 55 inches annually, compared to western states like California (22 inches) and Texas (27 inches) (NOAA 2015). Combined with an estimated 555 trillion gallons of water in aquifers (Hairston et al. 2023), Alabama has a significant water surplus. For comparison, Texans use an average of 16.5 million acre-feet (AF) of water a year (TAMU 2014), and Californians 72 million AF per year (DWR 2018). Alabamians use 9.23 million AF per year, with an estimated 1.6 billion AF in available groundwater and 102 million AF flowing through the rivers annually (ADECA 2015).

Of the 9.23 million AF, thermoelectric-power accounts for 80 percent of all water usage in the state, making Alabama Power (the state's major electric utility) the largest water consumer in the state. Although entirely supplied by surface water, over 80 percent of this water is returned to the original source following usage. Of the remaining 20 percent, almost half goes toward public supply of water, 31 percent to industrial use, with the remaining toward small segments such as irrigation and agriculture (ADECA 2018). Due to increasing efficiency with thermoelectric power, water usage has decreased 20 percent since 1980, despite a 25 percent increase in population. Additionally, due to the abundant surface water, only 6 percent of total water usage is groundwater, with the vast majority of that going to cities not located adjacent to surface water supplies (ADECA 2018). To put this water abundance into perspective, the lowest recorded flow of the Alabama River was 6.5 million AF in 2007; in its worst recorded year, Alabama's urban water consumption would only equal about 28 percent of just the Alabama River's lowest recorded flow, and its average annual consumption uses only about 2.3 percent of the state's surface water availability. Unlike the western states, even in its worst drought, Alabama collectively has rich water resources, which has shaped the exigency to which policymakers attach to water and drought policy within the state.

Due to the hydrological surplus, water is seen as an economic tool, meaning that it also hardly viewed as a scarce resource to be protected. As a state water researcher stated (Interviewee B 2022), "I don't think Alabama will get to a state water plan because we have plenty of water. There aren't a lot of negatives right now to not having a water plan because there are only a few years, maybe once out of every 20 years, we will need to curtail water usage." Instead, scientists like Interviewee B are working with environmental groups and major riparian owners to responsibly use more water to boost economic development, especially in the poorer parts of the state.

Water planning in Alabama is organized by the local watershed. Water distribution within the state is relatively uniform, with no real anthropogenic infrastructure to facilitate water movement across the state. The five major rivers in the state create stability with water supply and allow for abundant water resources throughout the state, as shown in Figure 5.1.

Alabama has groundwater, but accessing it is difficult, especially when drilling through limestone in the north (Interviewee G 2022). In the eastern part of the state, the groundwater is too deep to be economically feasible, while up north it is found in pockets, leading to situations where one town is in a water crisis, while another is relatively flush (Interviewee B 2022).

Structural Variables

Rules and Communication

The rules and structures of planning begin with state water law and then extend to the organizations charged with enforcement. Like most states in eastern United States, Alabama uses the doctrine of riparian rights (Andreen 2022). Based on English law, riparian rights allow landowners along the rivers (or surface waters) to access the resource, and in the case of Alabama, the laws

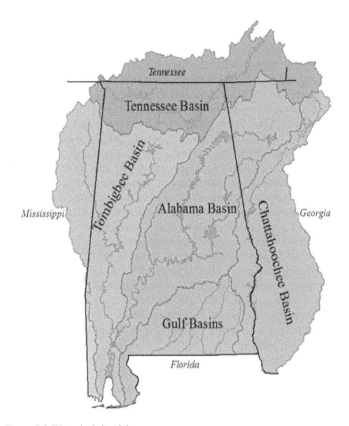

Figure 5.1 Watersheds in Alabama.

Source: Alabama Department of Economic and Community Affairs

can even restrict neighboring landowners from transporting the water across land, even with the riparian landowners' permission (Interviewee B 2022). Alabama also uses the legal concept of *beneficial use*, which it defines as "the diversion, withdrawal, or consumption of the water of the state in such a quantity as is necessary for economic and efficient utilization consistent with the interests of the state (Ala. Code 9–10B-3(2))." This means that a riparian user may consume water even if it could cause harm to a lower riparian user, unless that use is artificial. In these instances, riparian landowners may divert water for "artificial" purposes, such as electricity generation, but they must return the water to its natural source once complete (Andreen 2022).

Groundwater legal precedent in Alabama is largely based on mining and seeks to limit liability for incidental damages, except when subsidence damages occur to neighboring wells (Andreen 2022). Groundwater withdrawals are generally not regulated if for beneficial use; however, there are exceptions.

A case against the City of Linden (*Martin v. City of Linden1995)* was upheld after the government bought one acre of land miles away from the city and attempted to extract 500,000 gallons of water per day to be transported off-site for urban use. In this case, the injury caused to surrounding landowners was upheld due to the "reasonable use" of doctrine and removing the water from its natural source (Andreen 2022). Finally, due to navigational rights, the state owns all navigable river bottoms and tide shelves, although Alabama Code § 9–11–80 states that all waters are "public" waters (Andreen 2022). Moreover, the state Supreme Court has repeatedly refused to recognize "first-in-time" rights. In short, Alabama water law is relatively straightforward in usage rights and responsibilities. The abundance of water and established codes leave little conflict relative to water as compared to other more arid states. This provides a stable, albeit under-developed, legal precedent.

Water in Alabama is primarily managed by the Office of Water Resources (OWR) under the Department of Economic and Community Affairs (ADECA; see Table 5.1). Created by the Alabama Water Resources Act of 1993, OWR manages all water resources in the "best interests of the state" to include representing the state's interests in interstate negotiations (ADECA 2023). OWR views its primary mission as collecting data and producing reports on water usage, as well as participating in additional working groups and commissions; but as previously stated, due to the economic view of water, OWR falls under the Department of Economic and Community affairs, not the Department of Environmental Management or the Department of Conservation and Natural Resources (ADECA 2018).

Operating as part of the OWR, the Alabama Water Resources Commission consists of 19 members selected by the governor, lieutenant governor, and speaker of the Alabama House of Representatives. These members serve six-year terms and represent each congressional district in Alabama as well as each major water basin. Their primary purpose is to advise the governor and legislature by recommending policy changes, advise the Alabama Department of Economic and Community Affairs (ADECA) director, manage water rules and regulations, and oversee water enforcement within the state. Meetings from the Alabama Water Resources Commission are sporadic as they are required to meet only twice a year. There are no published meeting dates, and online records of the meetings are generally only one page of text with a basic outline of subjects with no details of the discussion (AWRC 2022).

Following a relatively severe drought a year prior, on April 16, 2008, the Alabama legislature created the Alabama Permanent Joint Legislative Committee on Water Policy and Management (Andreen 2022). The committee was charged with the task of "developing the Alabama Water Management Plan to recommend to the Governor and the Legislature courses of action to address the state's long-term and short-term water resource challenges (2008–164 SJR28)." The Committee served as an outlet for discussing water issues in the legislature; however, they have been unable to draft legislation for a state water plan.

With the emergence of another drought in 2011, Governor Robert Bentley attempted a different venue by creating the Alabama Water Agencies Working

Table 5.1 Water Management in Alabama

Office	Responsibilities
Alabama Office of Water Resources (OWR)	The agency responsible for the planning, coordination, development, and management of Alabama's water resources, in accordance with the Alabama Water Resources Act, coordinates efforts to compile and share information concerning hydrological data, water resource conditions, impacts, and drought mitigation responses by: 1. Coordinating the monitoring and collection of data; 2. Issuing Alabama Drought Declarations in accordance with this Plan; 3. Developing procedures necessary to collect and distribute information, convene committees, and promote water conservation and other means to encourage the wise stewardship of Alabama's water resources; 4. Coordinating and communicating Alabama drought declaration level and drought impact information; 5. Coordinating and communicating with the Alabama State Climatologist on recommendations and inputs; and 6. Encouraging the wise and efficient use of water.
Alabama Water Resources Commission (WRC)	The Alabama Water Resources Commission has both an administrative and advisory role in the overall management of Alabama's water resources, including state-level responses to drought conditions, in accordance with the Alabama Water Resources Act. 1. Provide advice and input to the Governor, Legislature, and AOWR in all aspects of drought planning, management, and response at the state level; 2. Approve any regulations proposed by OWR in support of this Plan; and 3. Approve any enforcement actions proposed by OWR.
Alabama Drought Assessment and Planning Team (ADAPT)	The Alabama Drought Assessment and Planning Team serves in an advisory capacity to OWR and the Governor's Office to coordinate intergovernmental drought assessments, responses, and management actions and in the implementation of all drought-related activities. 1. Provide guidance for various aspects of drought management; 2. Review the Alabama Drought Management Plan at least every five years to evaluate the performance and suitability of the drought indicators and the effect of pre-drought and drought responses; 3. Recommend appropriate changes; 4. Develop plans and procedures to support the implementation of a statewide drought planning and response processes; and 5. Provide guidance and make recommendations on drought-related matters to OWR and the governor, as necessary;

Group (AWAWG) to make updates for water programs and provide policy recommendations. This working group brought together five state agencies: The Office of Water Resources, the Department of Conservation and Natural Resources, the Geological Survey of Alabama, the Department of Environmental Management, and the Department of Agriculture and Industries. This group met over a two-year period, including three presentations for public comment, six regional water symposiums, and four publications (AWAWG 2013). Draft reports were sent to 248 water stakeholders, including environmental interest groups, academics, businesses, federal agencies. Direct feedback was returned by 82 of those individuals and groups and published online as well as categorized in the final report sent to the governor (AWAWG 2013). The report sent to Governor Bentley in 2013 called for an overhaul of the Alabama water system, including an analysis of current and future water needs, suggested revisions to Alabama water law, and a recommendation for a comprehensive state water plan (AWAWG 2013). Specifically, the working group advocated for an evaluation of the current riparian doctrine versus a regulated riparian model, as well as a method for local and regional input into water planning.

The AWAWG was reconvened in 2014 after the previous authorization expired. Again, they were tasked with engaging stakeholders in the effort to further inform a comprehensive state water plan. The AWAWG chairman, Dr. Berry H. Tew, Jr. briefed the working group's final recommendations to the new governor, Kay Ivey, on October 13, 2017. Following this meeting, the working group was disbanded due to the governor's view that "the waters of the state appear to meet both our current and future needs for some time to come" (Ivey 2017).

One of the first recommendations from the AWAWG was the creation of the Alabama Drought Assessment and Planning Team (ADAPT) as part of the Alabama Drought Planning and Response Act (Code of Ala. 1975, §§ 9–10C-1 et seq.). Chaired by the OWR division chief, this team consisted of experts throughout state government to provide interagency coordination for droughts in the state. Members included the Emergency Management Agency, Department of Agriculture and Industries, Department of Conservation and Natural Resources, Forestry Commission, State Climatologist, Watershed Management, and other appointed members and experts, who collectively advise the OWR and the governor on drought-related matters (ADECA 2018). ADAPT is vital to the state's water management since it is a standing committee, codified in the Drought Act of 2014. The final group that has input into water planning is the Environmental Management Commission under the Department of Environmental Management. This commission is focused primarily on environmental degradation across the state and is responsible for enforcement of water policy within the state. They regularly address water within the meetings but focus on environmental damage versus an overall management perspective (ADEM 2023).

Alabama does not have a state water plan. While lawmakers have pushed for a plan on several occasions, these attempts have each been unsuccessful. As seen earlier, water planning is shared among multiple departments, from economic development to environmental management. This creates a rather decentralized water planning process with overlapping responsibilities. While an overall state water plan is not available, the state does have a drought management plan (ADECA 2018). The drought management plan delineates the relationships of each of the agencies mentioned before, as well as lays the groundwork and foundation for coordination and communication across the state during periods of drought. The drought plan also divides the state into nine drought regions that allow for more nuanced input to the state, as shown in Figure 5.2.

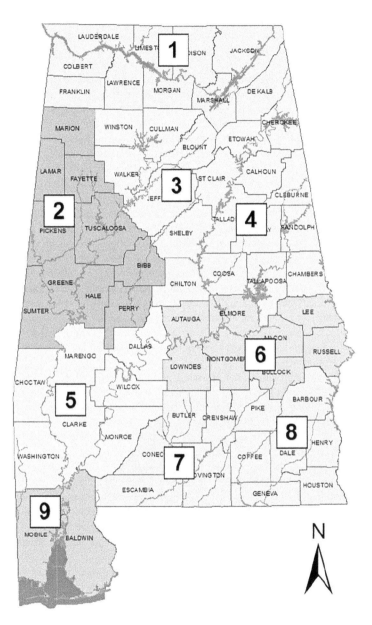

Figure 5.2 Alabama's Drought Management Regions.

Source: Alabama Department of Economic and Community Affairs 2017

Incentives

Incentives for water policy innovation vary by state and are often a result of grant opportunities and the availability of funding for water research. In the early 2010s, funding for environmental policy in Alabama was at an all-time low. In 2010, a group of 14 environmental organizations filed a petition to the Federal Environmental Protection Agency (EPA) seeking to turn evaluation and management of the state's environmental programs to the federal government as a result of a 72 percent cut in the ADEM budget (Woolbright 2016). After another decrease in the 2011 state budget, ADEM was forced to raise fee prices by 19 percent to mitigate the effects of diminished state funding (Associated Press 2015). After funds were cut again in 2013, ADEM raised its fees by 50 percent. Finally, in 2015 the legislature completely eliminated ADEM's general fund budget, with the exception of one small program, and began diverting $500k from ADEM's fees collection to other agencies. ADEM director Lance LeFleur cautioned the state that the EPA had threatened to rescind primacy for numerous environmental laws unless Alabama agreed to fund specific provisions of the Clean Air Act and Clean Water Act (Associated Press 2015). As was stated at an Environmental Management Committee meeting in October 2015, "The bottom line is we have virtually zero funding from the Legislature. So going with the fee increase, it does provide, presumably, more stable financing in that it will be there year after year." This sentiment closely mirrors the Alabama River Alliance's public comments that "we identified budgetary resources as one of the key problems we are facing in this state" (Associated Press 2015).

Due to public and federal pressures, lawmakers restored ADEM's general fund, currently listed as $4 million from 2019 to 2022. ADEM's remaining budget comes from fees, fines, and specifically funded programs. To put this into perspective, in the 2021 state budget (Alabama State Budget Office 2021), ADEM employed 648 personnel at a total expenditure of $50 million in personnel costs and benefits, but with $23M in federal grants and only $4 million from the state general fund. For Alabama, ADEM is effectively federally and self-funded. Additional funding for ADEM included a one-time $225 million in grant funds to upgrade water and sewer systems throughout the state, as part of Alabama's portion of COVID-19 relief money (ADEM 2022). State grants as part of the Alabama Irrigation Initiative have also become available to develop more sustainable agriculture throughout specific water basins (Davis 2020).

Federal funding plays a significant role in Alabama's water planning and policies. ADECA's website lists only one water-related grant opportunity, the Land and Water Conservation Fund, which is a federal fund available to communities who can match the funds requested. Any additional funding for water research must originate from a direct line-item request, as there are no existing grant programs within the state. While there are some grant opportunities from the influx of federal funds, incentives for water research and innovation are negligible. Even though the state is currently flush with funds, it is unlikely that this fiscal environment will continue. Due to previous droughts, the state has funded water working groups, but in the present environment, water planning and management do not appear to be top policy priorities.

Opening

Opening refers to the ability of citizen and interest groups to make their opinions known in state policy and to engage with policymakers. Citizen input to political processes in Alabama is mixed and is often dependent on the organization. Some organizations are willing to solicit citizen feedbacks, while others are comparatively opaque about their process as well as restrictive with information. The Alabama Resources Commission is the public face of the Office of Water Resources. Meeting times for the commission are not readily available, as they are internally decided (ADECA 2022). Only minutes for the last meeting are publicly available, and the minutes are on average a single page about the attendees and a single page of vague comments on topics discussed. Unless a citizen calls the OWR directly, there is limited information on past or future meetings.

Conversely, from 2011 to 2016, the Environmental Management Commission held 54 public meetings, an average of nine a year. Meetings are published months in advance, and 13 years' worth of archives are available online, including word-for-word transcripts and videos of each meeting. Agendas for each upcoming meeting are made public weeks in advance. Public comment at meetings is encouraged, including special meetings just for public comment on specified changes to environmental regulations, as well as online submission of comments.

Across the six years of this study, there were 26 individuals or organizations that participated in the public comment period (ADEM 2022). Certain topics garnered repeated interest, especially coal ash sewage around Uniontown, AL, in 2014, which comprised of half of the total participants of the six-year period. Besides the incident in Uniontown, the majority of inputs to the commission are various riverkeepers throughout the state. Of note, not a single national interest group participated in the Environmental Management Commission over the six-year period of this study.[1]

National interest groups, such as Sierra Club, have chapters in Alabama, but are not particularly active in the governmental processes, compared to states such as Texas and California. The most active water environmental group is the Alabama River Alliance, which consists of "riverkeepers" for most of the major river systems within the state (Lowry 2022). This group routinely writes opinion pieces published in state newspapers, attends state environmental meetings, and lobbies on behalf of the environmental needs of the state. Additionally, this group has been instrumental in pushing for a state water plan, although the current administration has halted progress on the legislation (Ivey 2017).

Resistance

Compared to more litigious states, such as California, there are relatively few environmental lawsuits in Alabama. Between 2011 and 2016, state courts did see an unusual spike in environmental lawsuits, with over 200 lawsuits filed, albeit many of these were filed against British Petroleum (BP), following the Deepwater Horizon oil spill in 2010 (Justia 2022). Excluding the BP lawsuits, stakeholders filed 37 environmental lawsuits within the state of Alabama during this period. Of these lawsuits, 15 of the 37 were filed by one of the various riverkeepers in Alabama, including the riverkeepers of Black Warrior, Tennessee, and Mobile Rivers, with nine of those cases against individuals or private companies. Only 4 of the 37 cases involved one of the state agencies as a plaintiff, suggesting a low rate of litigation in the state.

One of the primary hurdles to water/drought policy-related innovation is the lack of a common agreement on the problem, especially among environmental groups, regulated entities, and the primary water users. In Alabama, there is significant mistrust among the aforementioned groups. As Interviewee B (2022) stated, "if one side embraced it, the other side questioned why they should support it . . . they were afraid there was something they weren't seeing." Interviewee B continued that both sides have good reasons for mistrust; the environmental groups have at times taken positions that seemed extreme given the reality within the state, while major water users seemed uninterested in even considering a state water management plan. These large water users, such as the Alabama Farmers Association (Alfa) and Alabama Power, are content with the current water management and fear losing access to abundant and affordable water.

On the state level, the lack of cooperation has left lawmakers with little incentive to act. "If you know anything about the state, if you can't get Alfa and Alabama Power on board, it's probably not going anywhere" (Interviewee B 2022). But the mistrust goes deeper than just this one issue. Patrick McNally of the Department of Agriculture and Industries noted, "I've seen this change over the years. There is so much mistrust between the parties that in every legislation debate, even things we should agree on, everyone is suspicious. Everyone assumes there is something wrong with a piece of legislation if all the parties start to agree."

Another bureaucratic hurdle is a lack of state-appropriated environmental funds. The state simply refused to spend money on water planning or sustainment during the entire period of this chapter, although some of that is currently corrected. The environmental management offices have received consistent funding over the past four years, although funding to support innovative water solutions seems unlikely.

Cultural Variables

Shocks

Major shocks are often the spark that drives innovation in policy. Despite the relative abundance of water in Alabama, droughts have impacted citizens throughout the state. From July 2006 until February 2008, Alabama experienced its worst drought in over 50 years. Rainfall in 2007 measured 37.86 inches, well below the average of 55.25 inches. The only drier year on record is 1954, in which the state experienced 35.4 inches of rainfall (NOAA 2015). Drought-like conditions persisted for the next 18 months until heavy winter rainfall in 2008. Another drought began in 2011 and lasted until 2013. Although less severe than its predecessor, the combined drought in four out of seven years illustrated the necessity for drought planning and the dangers of drought impact within the state.

The difference between Alabama and most of the western United States is the lack of a catastrophic and long-term drought, and such droughts have impacted the population. Even in its worst drought period, water restrictions were mild as compared to western states. Periods of drought have brought water usage to the public consciousness, but they have yet to persist to the point of policy innovation. Moreover, periods of drought have led to discussion and the creation of new departments and working groups, but with a return to normal rainfall, many of these new groups were disbanded, and their recommendations were largely ignored (Ivey 2017).

Framing

One of the most difficult issues with environmental policy is defining the problem and framing it in a way that garners attention, builds supports, and secures the "frame's" place on the agenda. The Alabama Rivers Alliance (ARA) is the primary water advocacy group in Alabama, and Mitch Reid, the former program director, is seen as the voice of environmental action in the state (Alabama Rivers Alliance 2023b). Reid is consistent with framing the problem: Alabama's unique biodiversity and water management require statewide policy supervision. Alabama's biodiversity rivals the Amazon rainforest, with more species of fish, mussels, snails, and crawfish than in any other state and more species of turtles in the Mobile River system than in any river system on earth (Raines 2016). Reid is careful to focus on the problem, especially rallying points of pride for the state, which is further supported by ARA on its website. In an effort to bolster their coalition and framing, they have identified 39 other advocacy groups that support the ARA's water agenda and its framing of water.

Despite desires from environmental groups, both internal and external to the government, the state of Alabama largely dismisses calls for collective water planning. While droughts have increased state water research and information capacity, via the creation of groups like the Alabama Water

Agencies Working Group, subsequent budget cuts and years of water abundance stifled any momentum for innovation. Governor Kay Ivey noted, "The waters of the state appear to meet both current and future needs for some time to come (Ivey 2017)." So whereas state environmental agencies are collecting information and making recommendations on water management, the state legislature does not consider "water management" as a problem that warrants the investment into a statewide plan. As mentioned earlier in this chapter, one of the interesting twists on Alabama's water policy is the focus on economic development, especially in the poorer regions. Policy initiatives, such as the Alabama Irrigation Initiative, identified economic development as a precursor to encourage collective agreement vis-à-vis water management and planning (Interviewee B 2022; Lawrence 2021).

Increased Agenda Attention

As problems are framed appropriately, the increased public attention can be measured in news outlets and public interest. As a measure of agenda attention, news stories in the *Montgomery Advertiser*, the primary daily newspaper in the capital region, were analyzed based upon stories mentioning the term "drought." As drought conditions increased, there was a corresponding rise in news stories covering the said drought. Interestingly, the spike in stories before drought in late 2010 and the spike of stories from 2014 to 2016 correlate to peak drought conditions in both Texas and California, suggesting how news stories in one region can increase media attention in another. As local and national news sources increased attention to problems, Google searches by citizens also increased, as shown in Figure 5.3. The comparison

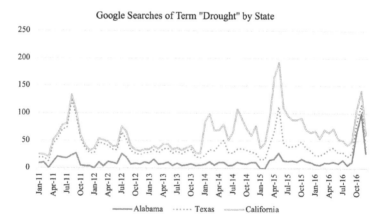

Figure 5.3 Google Searches of the Term "Drought" by State.
Source: Google Trends

of the data shows that on average, Alabamians appear less concerned about droughts overall than their western counterparts. This is indicative of the relative lack of concern of droughts in the region, where the agenda attention brought on by news coverage does not retain the public attention when the drought subsides.

Legitimacy

The state of Alabama does not appear concerned about legitimizing its position on water policy or the lack of a comprehensive water management plan. In 2017, OWR released the Alabama Surface Water Assessment Report, to which Governor Ivey (2017) released a statement saying that "even with population growth in the state's larger metropolitan areas and projected increases in industrial and agricultural use of water resources, Alabama should have a plentiful supply of surface water—rivers, streams, lakes and reservoirs—to meet those requirements through the year 2040." The state has repeatedly taken the position that no changes are needed to water management, which has similarly been echoed by major water users (ADECA 2017). The same press release continues by noting that OWR will begin to develop a statewide water management plan "if funded by the Legislature," but this appears quite unlikely. While the state did invest resources from 2017 to 2018 within the OWR and the Alabama Water Resources Commission to support the development of a state plan, research into a comprehensive plan was ended by Governor Ivey in 2019. In short, multiple working groups have investigated water in Alabama and have recommended a state water management plan; however, there has yet to be enough momentum to change the narrative within the state.

State environmental groups, including ARA, view their position through an environmental lens as compared to an economic one. ARA's website addresses economic development as an important use of water; however, repeated press releases focus on the environmental impact of water policy, which reflects their core mission as an environmental interest group. Groups like ARA are in a precarious position in the Deep South. They are very careful not to take strong political stances, such as is seen in other states like California. Instead, they leverage their bargaining power by focusing on individual issues, without usually naming specific politicians or political parties. Prior to a recent runoff election, ARA's executive director Cindy Lowry (2022) stated: "I can't tell you who to vote for—or even which party to vote for—but I can tell you that voting is very important to our mission of protecting the health of Alabama's rivers and people." With that said, ARA in particular has been vocal about the value of a state water plan. In conjunction with other environmental groups, such as the Southern Environmental Law Center, ARA has repeatedly criticized Governor Ivey for her decisions to disband the working groups. ARA's website (2023a) has multiple press releases on the water plan, as well as the recommended components of the plan once it is developed.

Policy Innovations, or Lack Thereof (2011–2016)

In 2014, the Alabama Legislature passed the Alabama Drought Planning and Response Act (Drought Act; Act 2014–400), the direct result of recommendations from the Water Agencies Working Group. This legislation directed the state to plan, monitor, and respond to drought conditions within the state. The act had two primary roles, to establish a structure for drought monitoring and to establish the state's role during a drought. The Alabama Drought Assessment and Planning Team (ADAPT) was also created and advises the governor and OWR on drought preparation and assessments (ADECA 2018).

There were no major legislation or policy updates following the 2016 drought. The current agencies continued their roles, but little debate occurred. The Department of Economic and Community Affairs has produced two drought management plans, the latest in 2018 (ADECA 2018). The drought management plan as written gives the state the responsibility to monitor drought conditions, but all drought mitigation and planning factors are expected to be held by community public water systems. Since most reservoirs are held federally through the U.S. Army Corps of Engineers or privately through the Alabama Power Company, the state has little jurisdiction over these resources (ADECA 2018). This means the state is primarily an information hub during the drought with little statutory power over the physical water within its borders.

Conclusion

This chapter was built to evaluate the four propositional statements outlined in Chapter 2. This conclusion will evaluate each proposition based upon the findings from this case.

(1) States with highly detailed, frequently reviewed water plans (<ten years) will be less likely to enact non-incremental policy innovations in response to a new drought event.

Alabama has yet to promulgate or implement a statewide water plan. The primary water users, such as Alfa and Alabama Power, are content with the status quo, so previous planning efforts have stalled before any plans reached the implementation stage. If Alabama can promulgate a plan, it will likely be an extension to the current policy. This case suggests a potential route to incremental (rather than innovative) policymaking.

(2) States with distributive water infrastructure will experience a more competitive and fractured water political culture.

Alabama does not have a distributive water system, with the only water infrastructure owned by the U.S. Army Corps of Engineers or by Alabama Power. There does appear to be a fractured water culture, but this was largely due to business interests and the lack of a major water quantity shock in the state.

(3) States with competitive environmental cultures will have less innovative policies.

Due to the lack of water quantity shocks, Alabama's "water management" culture is under-developed as compared to much of the country. With that said, debate over water policy was particularly contentious, with distrust between the Alabama Rivers Alliance and the large water businesses, primarily Alabama Power and Alfa, and lawmakers. The businesses even left the water negotiations due to concerns over environmental support. This distrust between the parties has effectively shuttered any hope of a water plan in Alabama. This case supports this statement.

(4) States whose citizens have experienced major water shocks will have a more cohesive, innovative policy culture.

While Alabama pushed for a state water plan in 2014, it was unable to overcome institutional resistance from major businesses and the state's political leanings. There was simply not enough public pressure to force action. After a worse drought in 2016, the state had no response at all. There has been no true shock severe (or long) enough to galvanize public opinion on water policy. Due to the lack of a true focusing event, Alabama's water culture was disjointed and incapable of enacting innovative policies. This case supports this statement.

Note

1 While ADECA and OWR are vague about citizen inputs and openness, the ADEM and the Environmental Management Commission utilize a citizen input plan. Unfortunately, for the purpose of this study, drought management is held under the umbrella of ADECA. While ADEM is responsible for water environmental standards and adherence, these tasks are separated from water management and drought planning. When asked for an interview, ADEM declined since they do not manage drought planning, but just environmental enforcement. ADECA and OWR also refused multiple interview requests.

References

Alabama Department of Economic and Community Affairs [ADECA]. 2018. *Alabama Drought Management Plan.* Montgomery: State of Alabama.

ADECA. 2015. "2015 Water Use Report—Fact Sheet." Accessed November 26, 2023. https://adeca.alabama.gov/wp-content/uploads/2015-Water-Use-Report-Fact-Sheet.pdf.

ADECA. 2017. "State Releases Comprehensive Assessment on Alabama's Surface Waters." Accessed November 26, 2023. www.media.alabama.gov/AgencyTemplates/adeca/adeca_pr6.aspx?id=12591.

ADECA. 2022. "Minutes." Accessed November 26, 2023. https://adeca.alabama.gov/wp-content/uploads/20221027-AWRC-Meeting-Summary.pdf.

ADECA. 2023. "Office of Water Resources." Accessed November 26, 2023. https://adeca.alabama.gov/water/.

ADEM. 2023. "The Environment Management Commission (EMC)." Accessed November 26, 2023. https://adem.alabama.gov/commission/default.cnt.

Alabama Department of Environmental Management [ADEM]. 2022. "Hundreds of Systems in Alabama Apply for $225 Million in Grants." *Alabama.gov.* March 29. Accessed August 31, 2022. https://adem.alabama.gov/newsEvents/pressreleases/2022/ADEMARPARelease32922FINAL.pdf

Alabama Department of Public Health [ADPH]. 2023. "Well Water." www.alabamapublichealth.gov/environmental/well-water.html#:~:text=Groundwater%20is%20the%20main%20source,as%20wells%2C%20for%20drinking%20water.

Alabama Rivers Alliance. 2023a. "About Alabama's Watersheds." https://alabamarivers.org/about-alabamas-rivers/.

Alabama Rivers Alliance. 2023b. "History." https://alabamarivers.org/history/.

Alabama State Budget Office. 2021. "Executive Budget Fiscal Year 2022." March. Accessed April 4, 2022. http://budget.alabama.gov/wp-content/uploads/2021/03/FINAL-State-of-Alabama-Budget-Document-FY22.pdf.

Alabama Water Agencies Working Group. 2012. "Water Management Issues in Alabama." http://alabamarivers.org/wp-content/uploads/2016/06/alabama-water-agencies-working-group-report-water-management-issues-in-alabama.pdf.

Alabama Water Agencies Working Group [AWAWG]. 2013. "Mapping the Future of Alabama Water Resources Management: Policy Options and Recommendations." https://adeca.alabama.gov/wp-content/uploads/December-2013-Alabama-Water-Resources-Management-Policy-Report.pdf.

Alabama Water Resources Commission [AWRC]. 2022. "Overview." https://adeca.alabama.gov/alabama-water-resources-commission/.

Andreen, William L. 2016. "Alabama Water Law." https://scholarship.law.ua.edu/cgi/viewcontent.cgi?article=1334&context=fac_working_papers.

Andreen, William L. 2022. 4 Waters and Water Rights AL-1-AL-52 (Amy Kelley ed., LexisNexis/Matthew Bender 2022). *U of Alabama Legal Studies*

Research Paper No. 4050035. https://ssrn.com/abstract=4050035 or http://dx.doi.org/10.2139/ssrn.4050035.

Associated Press. 2015. *Alabama Environmental Agency Votes to Raise Permit Fees After Budget Cut*. December 18. Accessed August 24, 2022. www.timesfreepress.com/news/2015/dec/18/alabama-environmental-agency-votes-raise-permit-fe/.

California Department of Water Resources (DWR). 2018. *California Water Plan Update 2018*. Accessed November 12, 2023. https://water.ca.gov/programs/california-water-plan/update-2018.

Davis, Debra. 2020. "Alabama Irrigation Initiative Funds Flowing in North Alabama." https://alfafarmers.org/alabama-irrigation-initiative-funds-flowing-in-north-alabama/.

Hairston, James E., Donn Rodekohr, Eve Brantley, and Mike Kensler. 2023. "Water Resources in Alabama." https://encyclopediaofalabama.org/article/water-resources-in-alabama/.

Interviewee B in discussion with the authors. May 19, 2022.

Interviewee G in discussion with the authors. May 4, 2022.

Ivey, Kay. 2017. "State Releases Comprehensive Assessment on Alabama's Surface Waters." *Alabama Department of Economic and Community Affairs*. December 11. Accessed September 7, 2022. https://adeca.alabama.gov/wp-content/uploads/2017-Surface-Water-Assessment-News-Release.pdf.

Jenkins, Clinton N., Kyle S. Van Houtan, Stuart L. Pimm, and Joseph O. Sexton. 2015. "US protected lands mismatch biodiversity priorities." *Proceedings of the National Academy of Sciences* 112 (16): 5081–5086.

Justia. 2022. *Justia Dockets and Filings*. Accessed September 5, 2022. https://dockets.justia.com/.

Lawrence, Maggie. 2021. *Alabama Extension Scientists Guide Smart Irrigation Expansion*. Accessed August 24, 2022. https://ocm.auburn.edu/research-magazine/2021-spring/articles/09-smart-irrigation-expansion.php.

Lowry, Cindy. 2022. "Alabama Rivers Alliance." *Why Do Elections Matter?* June 15. Accessed September 6, 2022. https://alabamarivers.org/why-do-elections-matter/.

Martin v. City of Linden, Alabama (667 So. 2d 732 (1995).

Morris, John C. 2022. *Clean Water Policy and State Choice: Promise and Performance in the Water Quality Act*. New York: Cambridge University Press.

Mueller, Randall. 2015. *Where to Find the Hunters, State by State*. November 9. Accessed September 9, 2022. https://business.realtree.com/business-blog/where-find-hunters-state-state.

NOAA. 2015. *Global Climate Report—Annual 2015*. Washington, DC: National Oceanic and Atmospheric Administration.

Pew Research Center. n.d. *Political Ideology by State*. Accessed September 8, 2022. www.pewresearch.org/religion/religious-landscape-study/compare/political-ideology/by/state/.

Raines, Ben. 2016. *Alabama's Rivers Are Crashing Thanks to Water Rules That Favor Industry*. November 15. Accessed August 24, 2022. www.al.com/news/mobile/2016/11/alabamas_rivers_are_crashing_t.html.

Texas A&M University [TAMU]. 2014. *Texas Water Law*. https://texaswater. tamu.edu/water-law.

Woolbright, Alexandra. 2016. "Budget Cuts Could Devastate Alabama's Already-Strapped Environmental Agency." *The Anniston Star*. August 1. Accessed August 29, 2022. www.annistonstar.com/news/budget-cuts-could-devastate-alabama-s-already-strapped-environmental-agency/ article_fa384252-38a0–11e5–8a4f-67f43e02f32c.html.

6 An Analysis of the Three Cases

This comparative research of innovative water policies illustrates the complexity of environmental policies in the United States. This book is based on a comparison of multiple states utilizing the policy innovation and implementation framework (Steelman 2010, 3), with a modification that operationalizes the framework variables across three levels: Cultural, structural, and environmental. Working from bottom up, environmental variables include the distribution of water in each state and the infrastructure in place to move that water. Structural variables include the rules and structured communication in place in the states, the incentives that encourage water innovation, opening within the policy process to external input, and finally resistance to innovation through litigation and state-level constraints. Finally, the cultural variables include the external shocks to the system, such as major drought events, the way different entities frame the problem, and how entities utilize narratives to legitimize their position. The combination of these bodies of literature creates one comprehensive framework that can account for the wide variation seen in policy innovation across those three levels: Cultural, structural, and environmental (see Figure 6.1).

The modified framework for this research provides a foundation for the analysis of water policy innovation from 2011 to 2016, from the path-dependent political structures to the environmental cultures shaped by years of competition and conflict. This chapter compares the findings of each state utilizing the framework variables, with a particular focus on the differences between the cases.

Environmental Variables

Water Distribution

The distribution of water refers to the amount and location of water resources in each state. This distribution has a significant impact on policy innovations in the selected cases. Alabama's annual 120 million acre-feet (AF) of surface water, evenly distributed throughout the state, stands in stark contrast to

DOI: 10.4324/9781003498537-6

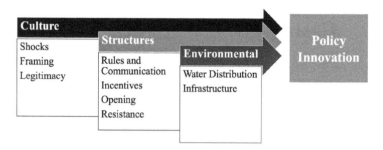

Figure 6.1 Policy Innovation and Implementation Framework With Environmental Variables.

Texas's 12.5 million AF of surface water, most of which is not located within 100 miles of a major city. In fact, if there are two opposites in water distribution, Texas and Alabama would be it. Texas is forced to account for every drop of water, through multiple emergency implementations of its drought plans (Interviewee P 2022). Although Alabama experienced periods of drought, the state has never faced the severity and duration required to be concerned about its water sources (Interviewee B 2022). By the time Alabama's political machinery was put into motion, the subsequent rains washed the problem away.

Somewhere between these two extremes lies California, with an average of 77 million AF of surface water, but unevenly distributed and serving a larger population base than the other states. While California has significantly more water than Texas, the water in California is geographically separated from the population centers. Where California differs from the other states is in the concentration of population and water usage. Texas and Alabama are relatively uniform in their population and power distribution. Alabama's major cities were distributed throughout the state, from Mobile to Huntsville, each with their own major water source. Texas similarly has most of its major cities in the center of the state with outliers like Houston and El Paso, but geographically deconflicted from each other for water.

California, on the other hand, has extreme differences in political power and environmental water distribution, as 69 percent of the population in the state is located within 30 miles from the coast (NOAA Office of Coastal Management 2023). Southern California is an engine for the U.S. economy, but it is hundreds of miles from its primary water sources in the Sierra Nevada and the Colorado River. For the California economy to function, water is taken from other regions of the state, or other states for that matter, and transported to the coast. This creates significant tension both internally and externally as more and more actors rely on the same water sources (Weible et al. 2011; Birkland 2006). Another unique characteristic of California is the relatively internal nature of its water. Except for the Colorado River, each of California's water sources is entirely internal to the state. This means that the state has relative carte blanche to move water as it sees fit, especially in the early days of California water, before the policy process was opened to external opposition (Birkland 2006). For Texas, the major rivers on the north and south each originate along the continental divide, whereas many of Alabama's rivers originate from the highlands of either Tennessee or Georgia.

Despite the shard nature of Alabama's water sources, the even population distribution and water abundance allow the state to absorb the shocks of droughts without requiring extensive water planning. While the state faces shortages of water periodically, drought does not capture the public attention and has not been worth the political capital required for policy innovations (Interviewee B 2022). This explains the relative lack of water innovation in the state, with a few interested actors, but not enough attention to put water policy on the state agenda.

Again, Texas could not be any more different from Alabama. The intensive water planning in the state is directly the result of its water distribution and shortcomings. Texas does not have the ability to absorb large shocks in the state and has proven to be even more susceptible to water fluctuations than previously thought. This drives a largely cooperative water culture, with ample funding and policy adaptations. One reason for cooperation is that the major cities are geographically distributed enough to not compete over the same water sources (Interviewee O 2022). In this sense, they are each in their own bubbles, relying upon regional water planning instead of competing over the same water. The only competition would be over state-level funding for water projects, but this was not a concern for any of the interviewees from the state.

Infrastructure

Infrastructure refers to the dams, locks, and aqueducts used to redistribute water throughout the states. The need for water stability along the arid California coast, combined with the almost unlimited economic funding of the region, drove the creation of some of the most ambitious water projects in the world. Unlike most urban planning, population centers in California were not limited in growth by their surrounding environment. This enabled incredible expansion in an arid desert, while creating a reliance on other regions for water. These major cities created water infrastructure well before the state formally managed it, meaning that the state largely inherited its water system from cities, such as San Francisco and Los Angeles, as well as the federal Central Valley Project (Interviewee M 2022). This led to a state-level management with inherited priorities of rights holders.

The redistributive water system in California created a series of "haves" and "have nots" that were based on political clout vice water availability. This ability to move water over large stretches of the state allowed Southern California to continue to grow and exert influence over less powerful regions of the state, which drove much of the current environmental conflict. While the mantra "farmers vs. fish" is used as a moniker for environmentalists living in Southern California to show the greed of farmers in the Central Valley, the truth is that the Central Valley has incredible natural water resources, such as the San Joaquin, American, Merced, Kings, and Kern rivers, but much of this water is moved out of the Valley to Southern California, leaving this region largely without its natural water. Unfortunately, Hollywood's image of sunny California is only sustained by the water from other parts of the state. Without the water redistribution, there would be no economic engine in Southern California, but there would also not be the level of conflict in the state.

On the other hand, Texas created a regionally based water system due to the lack of a large redistributive system. Texas does have over 200 reservoirs, but these are located close to the cities. Each major city began developing locally sourced water, which has then been combined into an overall state plan. Furthermore, the regionally based approach made each city responsible for its own water source development, instead of simply shipping water from other, less populated, regions. Alabama has even less infrastructure, with the majority of its lakes privately owned for power generation.

The key for these large infrastructure projects is that most were constructed before the "golden era" of environmentalism, beginning in the mid-1960s (Klyza and Sousa 2013). Up until this point, environmental activism was not organized on the same scale as today. The closed nature of the policy process and the lack of organized political opposition allowed massive projects that enabled the current disproportionate use of water in the state (Interviewee M 2022; Birkland 2006). The cost of infrastructure projects has increased 20-fold in the past 40 years, primarily due to environmental study and permitting (Interviewee O 2022). In short, the infrastructure projects that enabled California's development would simply not be affordable today, if the state were willing to pursue them. Even smaller proposals today, such as the Bay Tunnels (DWR 2022), have been mired in litigation with little movement in the past decade. Similarly, although Texas recently opened its first reservoir since the 1980s, the project took 20 years to permit. While infrastructure has had an impact on the resulting water policy, the status of that infrastructure has been largely static in the modern political environment and unlikely to change outside of a catastrophic event.

Structural Variables

Rules and Communication

Rules and communication refer to the water legal precedents in each state that have informed the state's institutional and organizational development. Texas and California represent two of the most developed water management systems in the United States but with markedly different processes. Texas's water management is reflective of its environment and traditionally conservative culture, with local-level planning and management that funnel up to a regional board, which then make recommendations to the state (Miller 2020). This structure is largely the result of the lack of water infrastructure combined with water planning around local water basins and aquifers. With that said, Texas invests heavily in water, a departure from a traditional conservative model, and represents one of the most progressive state water cultures (Interviewee P 2022).

Compared to this, California similarly invested heavily in water; however, the level of planning and focus was on the state level. Again, this was path-dependent since major cities like Los Angeles and San Francisco secured water sites hundreds of miles from their location well before the state began managing water (Interviewee M 2022). As such, water has been be managed on a state level to account for the widespread redistribution. Similarly, the federal construction of the Central Valley Project added an additional wrinkle with federal involvement in the state policies.

These two states have shared policy recommendations more frequently over the past two decades. Repeatedly interviewees from each state referenced the other, with a free flow of information over drought policies. Both states were happy with their current arrangement, to include differences, but the ability to consider other approaches and select the most appropriate option was evident.

Due to the lack of investment in water management, Alabama did not have the structured departments or processes seen in the other cases. There were overlapping organizations in the state with very little authority over water policy. Since Alabama did not have a water plan, there was no centralized structure to organize efforts. Until the state is able to set aside funding for a true water management body, it is unlikely to improve in the future.

This variable showed the impact of structured rules and communication on water policy. Since they had activated drought plans during these crises, interviewees in both California and Texas were very comfortable with the policy process, from long-term planning to crisis management. They understood their roles, and the structured process provided stability and predictability in the system (March and Olsen 1989, 23). Similarly, the implementation of drought efforts in the past allowed multiple services to be performed. First, it identified gaps in the knowledge and allowed for incremental improvement (Interviewee P2022). Second, activating plans allowed the "street-level bureaucrats" to work through the policy process, increasing their knowledge of how the process worked while simultaneously allowing feedback from the state to align their policy scripts (Weatherley and Lipsky 1977; Gioia and Sims Jr. 1986).

Incentives

Incentives refer to how each state encourages water innovation, primarily measured through water research and environmental funding. California is by far the most willing to fund environmental grants, with an annual average of $1.11 billion in water-based grants from the state. The state has a myriad of grant options and is heavily invested in water planning. Outside of pure funding, California leads the way in innovative ideas such as tunneling water through mountains and building hundreds of miles of aqueducts. The one limitation to this is that the desired innovation must match the state's vision for water management. As seen in the court case against the Westland's Water District, the state does not allow even self-funded research into increased water storage; this is incompatible with the state's water philosophy.

Texas law is more restrictive with grant opportunities, with only $172 million in water-based grants during the period of study, but the state often funds water improvements directly in the budget, investing $9 billion in water infrastructure upgrades since 2016. Similarly, major water projects are recommended by the regions but are often funded directly by the state. Both states have found different ways to fund water innovations but have been similarly effective.

Again, Alabama is the outlier here. For all practical purposes, the state has no grant funding for water innovation. The entire environmental budget for the state is currently $4 million annually (Alabama State Budget Office 2021). However, during the period of the study, funding had been reduced to zero, and some agencies were forced to turn their income from permits back to the state to fund other programs, forcing the EPA to threaten intervening in the state's environmental management. This is something inconceivable in the other states, as Texas and California are the leaders in water management in the United States. Although Texas has certainly been at odds with the EPA in the past, it has always funded environmental programs. The recent boon in Alabama water investment is only the result of federal financing and will likely fall off once that funding has been spent.

State funding of water policy is directly correlated to established policy innovations in these cases (Hopkins 2016). The reason that water practitioners in Texas and California are so comfortable in the process is that the processes and organizations are well funded. Each state has scores of people dedicated solely to water management, as opposed to Alabama, which is scarcely funded for its entire environmental management. Incentives in water policy innovation begin with funding. If states are unwilling to incentivize water policy, then they will have underdeveloped planning.

Opening

Opening refers to the ability for actors outside of the government, from citizens to NGOs, to provide input to water policy. Texas and California each had very open public involvement processes. Public comment was actively sought from local level to state, with repeated opportunities for public input.

Environmental groups participated in each process, with open dialogue between the state and interest groups as well as the incorporation of NGO input directly into long-term planning. Both states were praised for their open processes, so this is the strength of both systems.

Alabama's record during the period covered by this research was variable. Certain agencies were open to input, but others were not. Overall, the system was not actively seeking public input, but there is also little movement in water policy, so the effect of public input has been minimal. There was one instance of the water pollution in Uniontown, one of the poorest communities in the state, which was resolved following repeated public outcries. This illustrated that the community leaders had a proper venue that was open to their inputs, which in turn addressed their concerns. It's not that input was not accepted in Alabama, but rather that other than a few outliers, the public chose not to participate in the environmental management process.

The openness of these states points toward the success of their policies. Allowing input from outside voices enables a robust and adaptive process (Tyler and Moench 2012). The current strength of this approach stands in stark contrast to the early days of water policy in these states (Interviewee M 2022) but highlights how organizations that seek external feedback create hospitable environments for innovation (North 1990).

Resistance

Resistance within each of these cases refers to barriers or obstructions to policy innovation, from government red tape to recurring litigation. Of the three cases in this research, Texas is the more cooperative of the states, which is largely the result of the personal connection to recurring droughts. Texans have an acute sense of connection to the land and a clear understanding of its nature (Birkland 2006). While there has been some litigation in the state, the open communication between state agencies and environmental interest groups was evident based on published responses and interactions linked on the state websites. Most water-focused resistance in Texas occurs between cities over future reservoir sites (Interviewee O 2022; Interviewee C 2022). In this sense, it is less about what the current water should be used for, but rather how future water is obtained.

Alabama's resistance was seen more in the planning stages than in litigation. Large industries like Alabama Power and Alfa were diametrically opposed to environmental groups (Interviewee B 2022). If one group supported a policy, the other side became suspicious that something was amiss. Some of this breakdown was due to power imbalance at the table. The industry side had all of the power since they had legal precedent and land-ownership rights (Interviewee G 2022; Interviewee B 2022). Industry had nothing to gain from allowing more access to water and potentially something to lose if the state began managing water. Due to the inherent distrust of environmental groups in the state, progress on water policies effectively ended with industry once environmental groups supported water reform (Interviewee S 2022; Interviewee B 2022).

California was on the other extreme of resistance. Almost every water policy has been decided in the courts with each side fielding teams of lawyers and "water judges" controlling the state's policies (Interviewee H 2022). There have been periods of cooperation in the state, but competition was the most likely outcome. The use of voluntary agreements is a glimmer of hope in the future, but these agreements have not been tested by time, as previous agreements have not been successful in the long term (Interviewee M 2022). State agencies attempted to stay out of the conflict between environmentalists and farmers, although the annual allocations were the source of much discontent. Externally, the state has been contentious in sharing the Colorado River, not accepting a single reduction in its allotments despite dwindling reservoirs in the American Southwest.

One important difference between these cases was the amount of money involved. California spent billions of dollars on the environment, and each drop of water was directly tied to the economic potential of the farming lobby. In this case, the policy communities on each side of the problem were well developed and funded, leading to conflict over every detail (Birkland 2006). Each side was willing to invest in litigation to achieve its desired outcome, and often litigation in one area was used as a tool to extract a compromise in another (Interviewee R 2022). The actors in larger states, from major agricultural industry to environmental groups, used these legal victories as fundraisers for the next fight, so litigation became a staple of their business plans. Smaller states like Alabama were not as well developed on the state or the interest group level and could not compete with the resources dedicated in larger states. The policy communities were not as balanced, so the industry side had an advantage with organization and resources. Until industry has a reason to compromise, it is unlikely the status quo will change (Birkland 2006).

Cultural Variables

Shocks

Shocks refer to the major drought events that captured the public attention and raised awareness to potential water solutions. Texas and California each experienced shocks from previous water-focused events. Texas's drought of record is seared into the memory of every water planner, as well as the average citizen, almost reaching the point of folklore, and was referenced in every interview in the state (Interviewee P 2022). Similarly, California droughts in the past shaped surface water policy, but during the time of this study, the images of Californians walking miles for fresh water finally overcame institutional resistance to groundwater planning (Interviewee H 2022). The difference between these two states though was the primary purpose of the water in question. The conflict in Texas's policy was driven by cities trying to provide stable water for urban use. Conversely, California's water policy innovation was environmentally focused, but it did have somewhat hidden effects on smaller towns far removed from the major cities.

The difference between these states is that droughts have not been truly felt by the average Southern Californian. The water problems of dried-up lakes and land subsidence are hundreds of miles from the political power in the state. Other than increasing water prices, there is an insulation from many of the water problems caused by the droughts. These problems and fears are left for the junior water rights holders in the Central Valley to worry about. If the taps are still working on the coast, these problems will not be of their concern (Interviewee H 2022; Birkland 1997).

External to this research, but a key to comparative analysis, is the type of other disasters in the states. In Texas, water is one of the primary drivers of shocks for the state. There are other natural disasters, such as hurricanes and tornados, but water is the greatest concern. In California, droughts are just one of many concerns. From wildfires to mudslides and earthquakes, California has a bevy of natural disasters, many of which are a more imminent concern to the political class (Interviewee T 2022).

In contrast to each of these is Alabama, which has no true water-focused event in the past (Birkland 1997). While the state has experienced droughts, they have never been felt by the population and will likely not reach the level of public attention for some time (Downs 1972). In this sense, it is easy to understand why Texas is laser-focused on water management, California views water as just one of many environmental concerns, and Alabama is indifferent at best.

Framing

Framing is focused on the problem that each state is trying to solve and how each side of a potential policy issue used narratives to drive the conversations about the problem in their desired direction. The state of Texas frames water as a basic human necessity that needed to be addressed for future generations. Interest groups in the state include more of an environmental focus in their assessments, but they largely agree on a common definition of the problem, showing a largely coherent and collective understanding of the environment (North 1990; Gioia and Sims 1986).

In Alabama, the problem is framed as an economic issue, much more than environmental, with the state pushing for water policy in order to provide an economic boost for poorer, rural regions of the state (Interviewee B 2022; Interviewee S 2022). Unfortunately for Alabama, the state's framing of the problem does not match the views of powerful industries in the state, whose framing of the water surplus is able to override attempts at state-level water management.

California stands in contrast to each of these cases. State agencies walk a tightrope between environmental framing on one hand and economic growth in disadvantaged communities on the other. Water is most commonly framed by environmental groups as "farmers vs. fish" (Bonatakis 2019). The state struggles with this for two primary reasons. First, the state is a hub for environmental activism, strongly rooted in the culture along the coast and ideologically motivated (Besharov 2014; Shanahan et al. 2013). However, those regions are also the ones that need external water, so they are largely the benefactors of the water system that they loathe. Second, the Central Valley, which is the heart of the farming industry, is the most underserved area in the state with little political clout or influence. This is the region that has been largely left behind in California's economic boom. So, like Alabama, water in the Valley is exclusively an economic tool used to keep farming communities from spiraling further into poverty. Here the state is caught between two competing interests, allowing environmental flows that match underlying philosophy, while preventing abject poverty for an underserved population (Interviewee H 2022). As such, California takes the stance of co-equal priorities, allowing much of the conflict to take place between the actors themselves.

Legitimacy

In contrast to framing, legitimacy is focused on the potential solution to problems and how each side of the policy issue uses narratives to legitimize their potential solutions while delegitimizing their opponent's solutions (Shanahan et al. 2013). In Texas, due to the relative harmony on problem definition, both the state and environmental groups agree on conservation and infrastructure updates. The one divergence between these groups is the construction of more reservoirs, which the state sees as a necessity for future water security, whereas environmental groups believe infrastructure updates are sufficient to meet water goals (TWDB 2022).

Unlike Texas, Alabama's past attempts at water innovations were largely scuttled by the lack of cooperation between water agencies and industry. The state water agencies were unable to legitimize their positions on a statewide level that would allow for policy implementation. Distrust between environmental groups and industry was high, and the relative lack of environmental management from the state did not allow for movement on policy (Birkland 2006; North 1990).

During the period of this study, California took conflicting legitimacy of solutions to the extreme. Since the state only met its water allocation levels on average of one year out of ten, the state's water was truly a zero-sum game (Keohane 1984). There was simply no more water to be had, only water to be taken from one group and given to another. Unlike Texas, California was not interested in increasing water storage, so the only possible solutions were conservation and water reallocation (Interviewee R 2022). This was the heart of California's conflict and why this case was more conflictual than the other cases. The redistributive infrastructure allowed every drop of water to be a battle, from salmon hatcheries in the American River to artichokes in the Valley. This greatly increased the number of actors in the system and brought more groups into conflict (Sabatier and Weible 2007; Birkland 2006; Jenkins-Smith et al. 2014). Because of this, California had to balance a commitment to environmental ideals with the hard reality of providing water for millions of people and preventing the economic collapse of the world's food basket. This is accentuated by the over-allocation of water in the state, which helped explain the conflict between these groups.

The summary of key differences between the states is shown in Table 6.1.

Framework Evaluation

Steelman's (2010) Policy Innovation and Implementation Framework is particularly useful to analyze the varied angles of state water policies in these cases. Previously, the framework has been employed to investigate individual cases on the local and individual levels (Steelman 2010), but the utilization of this framework as a comparative tool should be considered for future research. The comparative operationalization of this framework necessitates the slight modification of the levels of analysis, pulling the focus from the individual to the state level. This study removes the individual variables of motivation, norms and harmony, and congruence due to the level of analysis. Adding the external environmental measures greatly increases the utility of the framework for this comparison and should be considered for future research.

The strengths of the framework are the multiple theoretical approach, from historical institutionalism to rational choice (Thelen 1999; Geddes 2003), the multi-level analysis, and the balance between simplicity and applicability. The framework as written provides explanatory power from the individual rational model all the way up to large cultural variability. The analysis across varying levels is particularly important to this research, as many of the cultural variables, such as shocks, directly affect the path-dependent structures of the states. Finally, the simplicity but utility of the framework strikes a balance for this project. Other environmental frameworks, such as the social-ecological systems (SES) framework, offer multi-tier analysis but do so at the expense of extreme complexity (Schlager and Cox 2017). Steelman's approach is multi-level but allows enough room for us to shape as required.

Table 6.1 Comparison of Cases by Variable

Independent Variable	Alabama	California	Texas
Environmental distribution	Average 120 M AF of surface water from 5 major rivers distributed across the state. Localized groundwater pockets in the state.	Average 77 M AF of surface water, heavily weighted to the north. Very little usable surface water in densely populated coastal areas. Overdrawn groundwater in Central Valley.	Average 12.5 M AF of surface water, unequally distributed on the northern and southern border. About 60 percent of all water use is groundwater, primarily weighted to high plains and hill country.
Infrastructure	Inconsequential water infrastructure in the state. Most dams and reservoirs are used for flood management or power generation.	Largest water distribution system in the United States consisting of federally owned Central Valley Project and state-owned California Aqueduct. Can move water from Lake Shasta in extreme north to San Diego on the southern border.	Over 200 reservoirs throughout the state. Most located within 100 miles of the desired metro area. Infrastructure is regionally focused. Unable to move water to large distances across the state. Major cities draw from different areas.
Shocks	Lacks the galvanizing focusing event to overcome opposition to water regulation.	Repeated droughts led to robust planning. The 2014 drought and subsidence led to SGMA policy innovations.	Repeated droughts led to robust planning. Each drought leads to reviews of assumptions and triggers.
Framing	Water is primarily framed as an economic potential, vice an environmental necessity.	"Farmers vs. fish" is the mantra of California. There is water, but the debate is between the most beneficial uses of it.	Some framing differences between environmental and state agencies, but not as extreme as California.
Legitimacy	If environmentalists agree, then industry is concerned that they must be missing something. Very distrustful of intentions.	Environmental groups are reticent to agree on water for anything other than environmental use. Agricultural districts point to their economic contribution to the poorest, underserved part of the state.	All sides agree on conservation and infrastructure improvements. State wants to build more reservoirs; environmentalists think conservation will suffice.

Independent Variable	Alabama	California	Texas
Rules and communication	Multiple overlapping agencies and authorities, but poorly funded. Very little coordination on the state level.	Very structured water planning process but tends to be top-heavy. Federal and state agencies dictate water allocations distributed to users. Can fluctuate wildly based on expected water availability. Senior rights holders never see water cuts.	Very structured water planning process, but planning is primarily on the local and regional level. State is used for prioritizing projects. Regionally focused due to dispersed water availability.
Incentives	Effectively no grant funding for water innovation. Entire environmental budget is $4 million annually. Current upgrades are due to federal funds.	Extensive funding for water innovation, average $1.11 B in water-based grants from the state.	Around $172 million in water-based grants during the period, but water infrastructure and sources are budgeted as line items vice grants. Around $9 B in water infrastructure upgrades since 2016.
Opening	Very little public involvement in water process. Some agencies were open and welcomed debate, while others were cryptic about meetings.	Very open public involvement process. Public comment sought from local level to state level, although state level is more shrouded in water allocation process.	Very open public involvement process. Public comment sought from local level to state level. Open dialogue between state and environmental groups.
Resistance	Resistance primarily from industry. Little litigation, but no interest from the state in developing water plans.	Water policy is decided in the courts. Almost every decision is litigated, although recent use of voluntary agreements hopes to decrease this in the future. State tries to walk a line between environmentalists and farmers.	Some disagreement between environmental groups and state, mainly over reservoir construction. Mutual agreement on conservation and infrastructure improvements.

Grounded Theory Implications

The primary themes identified by interviewees were the direction of drought planning, the importance of shocks on the policy process, the impact of water distribution to planners, the importance of citizen input to spur policymakers, cooperation and conflict in the environmental cultures, planned adaptation due to the changing environment, and the diffusion of policies across multiple states. While many of these themes were present in the framework, two stood out as an addition to theory: water access and planned adaptation.

Access to water, captured through the addition of the *environmental distribution* variable, was repeatedly referenced by interviewees as vital to policy outcomes. The relative deconfliction between water sources in Texas led to the local and regional focus in water planning and alleviated much of the conflict commonly found in environmental policy. Conversely, the massive infrastructure and long-distance transfer of water in California necessitated state-level management and drove litigation over water from San Francisco to San Diego. This one theme explains most of the variation between California and Texas and should be considered foundational to future water research.

The second additional theme is planned adaptation, which is largely undeveloped in environmental literature. Planned adaptation is expensive and requires a heavy investment in continual research and updates to disaster plans. However, every interviewee stressed how much plans changed and morphed as new information was assimilated. The most basic of assumptions, from population growth to aquifer sustainability, were tweaked with every drought event. Without these updates, the states could not have handled successive droughts. Unfortunately, it is difficult to enumerate this investment, as disasters avoided are an impossible metric to quantify. States like California and Texas have learned the hard lessons of the past and are willing to invest whatever is required to prevent the next drought disaster.

Conclusion

The variables used in this analysis each had an impact on at least one of the cases and worked together to explain policy innovation in Texas and California and a lack of innovation in Alabama. It would be impossible to reduce these complex problems to a single measure, or even a handful of measures. Environmental problems and policy innovations are complex and extend well beyond simple measures such as political ideology or regional location (Sabatier and Weible 2007). The importance of the environmental distribution of resources has largely been ignored in the literature up to this point, but its inclusion helps explain much of the variance seen in these cases and points toward its importance moving forward.

Collectively, the measures of this research framework paint a picture of the complexity of water policy and the importance of states adapting to multiple shocks, as new data becomes available. Water policy is not static. Due to the constant flux of the climate, states that invest in water research are the most prepared for the resultant crises. It is this recurring adaptation that makes for resilient water policy.

References

Alabama State Budget Office. 2021. "Executive Budget Fiscal Year 2022." March. Accessed April 4, 2022. http://budget.alabama.gov/wp-content/uploads/2021/03/FINAL-State-of-Alabama-Budget-Document-FY22.pdf.

Besharov, Marya L. 2014. "The Relational Ecology of Identification: How Organizational Identification Emerges when Individuals hold Divergent Values." *Academy of Management Journal* 57 (5): 1485–1512.

Birkland, Thomas A. 1997. *After Disaster: Agenda Setting, Public Policy, and Focusing Events*. Washington, DC: Georgetown University Press.

Birkland, Thomas A. 2006. *Lessons of Disaster: Policy Change after Catastrophic Events*. Washington, DC: Georgetown University Press.

Bonatakis, Lauren. 2019. *Farmers vs. Fish: The Story of Delta Smelt*. June 24. https://envirobites.org/2019/06/24/farmers-vs-fish-the-story-of-delta-smelt/.

California Department of Water Resources (DWR). 2022. "Delta Conveyance." Accessed October 10, 2022. https://water.ca.gov/Programs/State-Water-Project/Delta-Conveyance.

Downs, Anthony. 1972. *Inside Bureaucracy*. Boston: Little, Brown.

Geddes, Barbara. 2003. *Paradigms and Sand Castles*. Ann Arbor: The University of Michigan Press.

Gioia, Dennis A., and Henry P. Sims Jr. 1986. *The Thinking Organization*. New York: Jossey-Bass Inc.

Hopkins, V. 2016. "Institutions, Incentives, and Policy Entrepreneurship." *Policy Studies Journal* 44: 332–348. https://doi.org/10.1111/psj.12132.

Interviewee B in discussion with the authors. May 19, 2022.

Interviewee C in discussion with the authors. July 29, 2022.

Interviewee G in discussion with the authors. May 4, 2022.

Interviewee H in discussion with the authors. May 20, 2022.

Interviewee M in discussion with the authors. August 4, 2022.

Interviewee O in discussion with the authors. July 25, 2022.

Interviewee P in discussion with the authors. October 19, 2022.

Interviewee R in discussion with the authors. August 31, 2022.

Interviewee S in discussion with the authors. May 4, 2022.

Interviewee T in discussion with the authors. July 27, 2022.

Jenkins-Smith, Hank C., Daniel Nohrstedt, Christopher M. Weible, and Paul A. Sabatier. 2014. "The Advocacy Coalition Framework: Foundations, Evolution, and Ongoing Research." In *Theories of the Policy Process*, edited by Paul A. Sabatier, 183–224. Boulder, CO: Westview Press.

Keohane, Robert O. 1984. *After Hegemony: Cooperation and Discord in the World Political Economy*. Princeton: Princeton University Press.

Klyza, Christopher McGrory, and David J. Sousa. 2013. *American Environmental Policy: Beyond Gridlock*. Cambridge, MA: MIT Press.

March, James G., and Johan P. Olsen. 1989. *Rediscovering Institutions: The Organizational Basis of Politics*. New York: Free Press.

Miller, Kenneth. 2020. *Texas vs. California: A History of their Struggle for the Future of America*. New York: Oxford University Press.

NOAA Office for Coastal Management. 2023. "Understanding and Planning for Sea Level Rise in California." https://coast.noaa.gov/digitalcoast/stories/ca-slr.html.

North, Douglass C. 1990. *Institutions, Institutional Change, and Economic Performance*. New York: Cambridge University Press.

Sabatier, Paul A. and Christopher M. Weible. 2007. "The Advocacy Coalition Framework: Innovations and Clarifications." In *Theories of the Policy Process*, edited by Paul A Sabatier, 189–220. Boulder, CO: Westview Press.

Schlager, Edella, and Michael Cox. 2017. "The IAD Framework and the SES Framework: An Introduction and Assessment of the Ostrom Workshop Frameworks." In *Theories of the Policy Process*, edited by Christopher Weible and Paul A. Sabatier. New York: Westview Press.

Shanahan, Elizabeth A., Michael Daniel Jones, Mark K. McBeth, and Ross R. Lane. 2013. "An Angel on the Wind: How Heroic Policy Narratives Shape Policy Realities." *Policy Studies Journal* 41 (3): 453–483.

Steelman, Toddi A. 2010. *Implementing Innovation: Fostering Enduring Change in Environmental and Natural Resource Governance*. Washington, DC: Georgetown University Press.

Tyler, Stephen, and Marcus Moench. 2012. "A Framework for Urban Climate Resilience." *Climate and Development* 4 (4): 311–326.

Texas Water Development Board (TWDB). 2022. *2022 State Water Plan*. Austin: Texas Water Development Board.

Thelen, Kathleen. 1999. "Historical Institutionalism in Comparative Politics." *Annual Review of Political Science* 2 (1): 369–404.

Weatherley, Richard, and Michael Lipsky. 1977. "Street-Level Bureaucrats and Institutional Innovation: Implementing Special-Education Reform." *Harvard Educational Review* 47 (2): 171–197.

Weible, Christopher M., Paul A. Sabatier, Hank C. Jenkins-Smith, Daniel Nohrstedt, Adam Douglas Henry and Peter DeLeon. 2011. "A Quarter Century of the Advocacy Coalition Framework: An Introduction to the Special Issue." *Policy Studies Journal* 39 (3): 349–360.

7 Conclusions and Future Directions

The centrality of water to support human life focuses our attention squarely on the question of how best to manage water resources, especially during periods of scarcity. For decades, citizens and governments have developed patterns of water usage and built an infrastructure to exploit that water, which rely on a stable climate and weather patterns. However, a prolonged period of climate change upended many of the assumptions about water upon which this infrastructure has been built. In areas in which water is scarcer, this has the effect of raising the level of conflict present in a struggle to secure enough water to maintain existing levels of use. How states are able to respond to these changing conditions is the focus of this study.

This research utilized three case studies to test the importance of recurring shocks, planned adaptation, and competitive environmental cultures on drought policy innovation. Four propositional statements were presented as a foundation for this study and were evaluated throughout this project.

(1) States with highly detailed, frequently reviewed water plans (<ten years) will be less likely to enact non-incremental policy innovations in response to a new drought event.

California and Texas each showed the importance of frequently reviewed water plans. Both states had five-year planning cycles with well-funded water research bodies. After major droughts between 2011 and 2016, both states produced incremental updates to their respective state water plans. However, there is one importance difference between the states' water planning up to that point: groundwater.

Groundwater in Texas, although not controlled directly on the state level, has been enforced through a growing number of groundwater districts for decades. While these are locally and regionally governed, the state has repeatedly encouraged these districts. One of the reasons for this is the fact that much of the state relies upon groundwater for urban use, especially in the arid panhandle and south around San Antonio. Groundwater usage has been a part of overall water planning since the 1950s.

DOI: 10.4324/9781003498537-7

California's groundwater use is primarily focused on the sparsely populated, but agricultural dominant, Central Valley. Urban water outside of the Valley is almost exclusively derived from surface waters, through the Central Valley Project and the California Aqueduct. As such, surface water is highly developed, while groundwater is not. Although SGMA is a major innovation in water policy, it still affects a small percentage of California's population. Given the inclusion of groundwater in planning documents following the passage of the SGMA bills, this major shift in policy should help stabilize water in the future.

The Alabama case argued this point from the other direction. Without any effective water planning institutions, entire new working groups were established after major droughts, which took time to organize before they could make recommendations. Although they were disbanded prior to any policy innovations, any changes made to Alabama law would have been a large punctuation since water was (and still is) largely unregulated.

The cases in California and Texas both support this statement. Despite the lack of previous groundwater management, California was able to learn valuable lessons from states like Texas and incorporate a refined policy from the start. These states implemented large punctuations in water policy in the past, but with successive reviews, these policies became more stable with incremental improvements. Alabama was neutral on this statement since it had no appreciable innovations; however, if there is a focusing event in the future, it will likely produce a large departure from current water policies.

(2) States with distributive water infrastructure will experience a more competitive and fractured water political culture.

California's traditional water culture was the very definition of competitive and fractured, as litigation has been the status quo in California for decades. The water system in California was perpetually over-allocated, so water was a zero-sum game in the state. The ability to transfer water hundreds of miles to millions of people was a great blessing for California, but it was also a curse. While the state made great strides in voluntary agreements, the longevity of these agreements has yet to be tested.

Texas and Alabama did not have the same distributive water system. Their distribution systems were based upon regional water resources, from rivers to aqueducts. While there were reservoirs and canals for watering the cities in Texas, the major urban areas were geographically distributed throughout the state. Dallas was not competing for the same water resources as Austin or San Antonio. This deconfliction led to cooperation and stability in planning. As the population grew, these cities were each looking to expand water resources from their immediate east, but they were not vying for the same sources.

In Alabama, the major players for water were Alabama Power and Alfa, but with such a water abundance, the only conflict was over the environmental impact of river flows and effects of dams on fish populations. The riparian legal structure of the state gave power exclusively to the landowners, which removed much of the competition over water.

These case studies supported this proposition, although future analysis of long-term stability could alter these findings. As an example, California used litigation heavily in water policy, but litigation was also a key in almost every other environmental field. The litigation that appeared in this case could also be the result of a larger environmental culture, where groups view courts as a normal part of the policy process, rather than a means to address major disagreements. If California stabilizes its long-term water planning through voluntary agreements, followed by a decrease in water litigation, then this case could provide a better test for this statement. As it is, these cases support this statement, but not strongly.

(3) States with competitive environmental cultures will have less innovative policies.

California is the only state in this study that meets the threshold for competitive environmental cultures. Interestingly though, California manages to exhibit both a highly competitive environmental culture and innovative policies. One of the reasons for this, somewhat outside of this research, is that the state is effectively a one-party state, with the large farming industry represented by a very small minority in the state legislature. As such, policy on the state level is relatively stable, with the swings in water policy mostly felt by a small percentage of the population, far removed from the political power along the coast. Due to these contradictions in culture and policy, this case study does not support this statement.

(4) States whose citizens have experienced major water shocks will have a more cohesive, innovative policy culture.

California and Texas each experienced major shocks in water planning. While both states felt the impact of droughts, the population of California was largely insulated from this reality since the major shortages in water were felt primarily by farmers in the Central Valley. With 69 percent of the population living on the coast, Californians did not live with their water supplies, which removed citizens from the consequences of their water policies. There is a long history of taking water from other parts of the state to feed the thirsty coastlines. From Owens Lakes and the Hetch Hetchy Reservoir to the Colorado River, the need for water on the California coast destroyed many other ecosystems in the state, while leaving the coasts artificially green.

While Texas is far from perfect in water management, each city is essentially collocated with its water sources. There is much more of a tie between citizens and the water in Texas than is seen in California. Furthermore, Texas has less surplus in its water planning, so successive droughts have more of an impact on citizens. There simply is not as much water to go around, so the cities and regions must cooperate to provide basic services. Again, the major cities will be a top priority based on water projects within the State Water Plan, but successive droughts have created a more cohesive water culture in the state.

Finally, Alabama has not experienced the true drought disaster required to create a cohesive water culture. There is no reason for industry to participate in water reform because there is plenty of water to go around, and the citizens have not demanded participation. Until there is a crisis that drives public attention, the water culture will remain fractured, and the policies will remain underdeveloped.

Overall, these cases support this statement. While citizens in California were aware of drought conditions, the only real fear on the coast was rate increases. Texans created a more cooperative water community and were tied more closely to their land and water sources.

Implications for the Study of Environmental Policy

This research focuses on several topics important in the larger field of environmental policy. One of the defining characteristics of the field of environmental policy, and specifically collective action, is the complexity of these problems and potential solutions (Olson 1965; Ostrom 2010; 1990). This research specifically addresses the literature of socio-environmental systems (Schlager and Cox 2017), path-dependent water policies (Pierson 2004), and focusing events (Birkland 1997). Across the board, socio-environmental systems research is messy and complex (McGinnis and Ostrom 2014). It cannot be reduced to a few simple measures that explain the variability in cases. As such, this research contributes by showing how two states, Texas and California, each took different paths and policies to arrive at a very similar destination. Each state adjusted to its own environments to create complimentary but distinct policies.

Outside of its environment, each state in this research is bound by prior legal precedents and pre-existing water policies, creating a path-dependent outcome (Pierson 2004, 20). Unique examples from these cases that still affect current policy in the states include California water rights that extend back to the Spanish governors, Texas's legal precedent that does not recognize "beneficial use" but does encourage regional groundwater management, and Alabama's code that prevents the surface transportation of water across private property, even if both parties agree.

Similarly, early decisions in the formative stage of policies have had an outsized effect on the outcomes (Pierson 2004, 45). The construction of large redistribution irrigation by individual cities in California established this policy as the norm by the time the state codified water management. Likewise, the privatization of water bodies by industry in Alabama largely removed the state from management. These cases each point to the importance of early, often unintentional, decisions on future policy options.

Finally, each of these cases points to the critical role of shocks, such as focusing events (Birkland 1997). California and Texas have been forced to respond to extreme drought events that raised the concern of drought to the public attention. As a result, they formulated varying levels of water policies based on the public appetite and necessity. These cases clearly add to the focusing event literature, both in the necessity and in the threshold of event required to spark a response.

Future Framework Application

Steelman's policy innovation and implementation framework provides a valuable method for analyzing environmental policy. As written, it provides robust explanatory power for single cases, but as a comparative tool, the inclusion of environmental variables increases the framework's utility. The variables added in this research, water distribution and infrastructure, are perhaps too narrow for the larger field of environmental policy. A more relevant approach may be adding *Environmental Distribution*, operationalized similarly to this project, and *Environmental Risks*. Environmental distribution could refer to resources, such as oil and natural gas, or location of forests in a region. Environmental risks could be operationalized as the propensity of environmental danger in a region, such as recurring droughts, earthquakes, or susceptibility to pollution. An example of this would be how the smog-prone Los Angeles Basin led to California's strict car emissions policies. These variables greatly increase the validity of the framework and should be considered for future research.

Areas for Future Research

Planned adaptation is a concept that is not well developed in most policy fields, but a cornerstone of real-world environmental policy. Future studies should investigate the link between iterative water plans and resilience to emerging shocks. As an example, one flooding expert mentioned how Louisiana has developed robust flood planning in the wake of Hurricane Katrina in the past 10 years and is now on the third iteration of its plan (Interviewee T 2022). California subsequently borrowed the post-Katrina water planning in Louisiana as the foundation for its own planning. Due to California's willingness to invest heavily in water planning, it has been able to take other lessons learned from other states and implement them before facing the disaster itself.

Another point of study would be why states are willing to invest heavily in certain aspects of water planning, while ignoring others. From our previous example, why did California adopt flood planning in the immediate wake of

Katrina, but not address groundwater until subsidence began destroying its infrastructure? Or, why has California over-allocated water so that it can only meet promised demands one out of every ten years?

An additional important topic for research would be the endurance of water agreements during the next major shock. California's use of voluntary agreements is an important step forward for the state; however, it will require time to see if these agreements hold. Previous agreements, such as the Monterrey Accords, did not survive successive droughts, so it is too soon to tell whether these policies are revolutionary or just a short-term fix. Furthermore, groundwater management in the state will require future research as these policies take hold.

Similarly, interstate policies around contentious water basins will require further research, such as the Colorado and Chattahoochee Rivers. This research is almost exclusively focused on state policies; however, the conflict between states is equally indicative of the water culture. More research will be needed to compare these intrastate and interstate cultures, especially given the movement of water hundreds of miles from the shared source.

Another topic of study was recommended by the eastern U.S. water planners. Experts within the water field have expressed concern about the large-scale redistribution policies that create false, fragile environments. As an example, modern farming has artificially concentrated crops in regions, such as the fruit basket in Central California or the breadbasket in the Midwest. Concentration of crops in relatively small regions has created instability in U.S. food markets, as an intense regional drought can now decimate the entire production of a specific crop. The ability to distribute crops across the larger growing climates may slightly increase the cost of crops but would provide more stability to the markets.

Closely related to this topic is the unequal federal farming subsidies. This idea of "crop migration" toward these super areas of production has been largely the result of U.S. government policies. As an example, improvements to the Mississippi River system by the U.S. Army Corps of Engineers allow relatively inexpensive grain transport down the Mississippi River (Interviewee B 2022; see also USACE 2009), which has lowered the cost of agriculture in the Midwest. Although intended as a post-dust bowl economic boost, the policy is still in place. As a result, other agrarian economies cannot compete with Midwest grain due to locally focused federal policies. Similarly, the U.S. Bureau of Reclamation heavily subsidizes irrigation projects, such as those seen throughout California. The downside is that the Bureau is only authorized to operate in the 17 western states (USBR 2023), meaning that other regions of the country that could utilize irrigation to grow their economies are not legally allowed to apply for funding.

Additionally, there is concern that the initial influx of water that filled western reservoirs such as Lake Meade allowed for population growth well beyond the natural environmental capacity. As this situation continues to develop, researchers should investigate the fragility of these environments that were artificially created in the American Southwest. Finally, each successive drought tests these state water plans. As the states continue to innovate water policies, revisiting these propositional statements would allow even more insight to incremental changes and planned adaptation. How states respond to the new policy imperatives will speak volumes about the ability of governments to provide water to meet their diverse needs.

Conclusion

These three cases demonstrate the impact of multiple variables on water policy innovations. As water is stressed more and more in the future, the ability to draw on lessons learned by these states will provide valuable insight to other entities that face droughts and water shortages. Repeatedly interviewees discussed diffusing policy ideas from other states. These cases show two vastly different approaches, based upon their respective environments. As states are looking for water management techniques in the future, understanding the factors that led to different but successful water management solutions will be invaluable.

References

Birkland, Thomas A. 1997. *After Disaster: Agenda Setting, Public Policy, and Focusing Events*. Washington, DC: Georgetown University Press.

McGinnis, Michael D., and Elinor Ostrom. 2014. "Social-Ecological System Framework: Initial Changes and Continuing Challenges." *Ecology and Society* 19 (2): 30.

Olson Jr., Mancur. 1965. *The Logic of Collection Action*. Boston: Harvard University Press.

Ostrom, Elinor. 1990. *Governing the Commons*. Cambridge: Cambridge University Press.

Ostrom, Elinor. 2010. Analyzing collective action. *Agricultural Economics* 41: 155–166.

Pierson, Paul. 2004. *Politics in Time: History, Institutions, and Social Analysis*. Princeton: Princeton University Press.

Schlager, Edella, and Michae Cox. 2017. "The IAD Framework and the SES Framework: An Introduction and Assessment of the Ostrom Workshop Frameworks." In *Theories of the Policy Process*, edited by Christopher Weible and Paul A. Sabatier. New York: Westview Press.

United States Army Corps of Engineers (USACE). 2009. "Inland Waterway Navigation: Value to the Nation." *USACE Institute for Water Resources*. Accessed December 8, 2023. www.mvk.usace.army.mil/Portals/58/docs/ PP/ValueToTheNation/VTNInlandNav.pdf

United States Bureau of Reclamation (USBR). 2023. "About us—Mission." Accessed November 12, 2023. www.usbr.gov/main/about/mission.html.

Appendix
Interview Protocol

1. What is your current position?
2. What is the scope of your responsibilities?
3. How long have you been in your current job?
4. Have you held any other positions in water policy?
5. How does your organization fit within the water policy process?
6. How often does your organization review existing water plans or policies?
7. How have previous droughts informed your organization's current thinking, policies, or goals vis-à-vis droughts?
8. What policy innovations have you identified but been unable to implement?
9. What hurdles or opposition prevented this policy innovation?
10. How do citizen/interest groups interact with your organization relative to droughts?
11. (California respondents) How did you incentivize the Sustainable Groundwater Management Act?
    (Texas respondents) How did you incentivize water reforms, such as Senate Bill 332 (2011)?
    (Alabama respondents) Why do you think Alabama has been unable to implement a state water plan?
12. Would you describe the water community in your state as cooperative, or is there conflict between different organizations or interest groups?
13. If conflict exists, what do you think is the reason for conflict?
14. How has the distribution of water throughout the state affected water policy?
15. How does your state utilize federal funding?
16. Does federal funding play a role in state water policy?

Index

Note: Page numbers in *italics* indicate a figure and page numbers in **bold** indicate a table on the corresponding page.

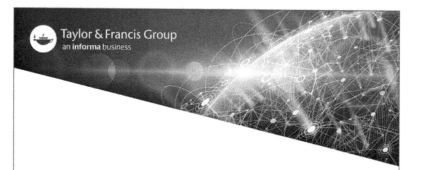

Milton Keynes UK
Ingram Content Group UK Ltd.
UKHW031137141024
449569UK00006B/119